Astronomers' Universe

Other titles in this series

Origins: How the Planets, Stars, Galaxies,
and the Universe Began
Steve Eales

Calibrating the Cosmos: How Cosmology Explains Our
Big Bang Universe
Frank Levin

The Future of the Universe
A. J. Meadows

It's Only Rocket Science: An Introduction to Space
Enthusiasts (forthcoming)
Lucy Rogers

Martin Beech

Rejuvenating the Sun and Avoiding Other Global Catastrophes

 Springer

Martin Beech
Astronomy Department
Campion College
The University of Regina
Regina, SK, Canada S4S 0A2
beechm@uregina.ca

Library of Congress Control Number: 2007934755

ISBN: 978-0-387-68128-3   e-ISBN: 978-0-387-68129-0

Printed on acid-free paper.

Printed in the United States of America.

9 8 7 6 5 4 3 2 1

springer.com

This book is dedicated to Georgette
my past, present, and future.

About the Author

Dr. Martin Beech is an associate professor of astronomy and head of the Astronomy Department at Campion College, The University of Regina in Canada. His main research interests during the past decade have focused on the smaller objects within the Solar System (comets, asteroids, and meteoroids), but concomitant to this he has continued to perform research related to the structure and evolution of stars (the area of his doctoral studies). The material in this book was partly based on a series of research papers Dr. Beech has had published in scientific journals, and the topic was the focus of an article written for the May 2006 issue of *Astronomy Now* magazine.

Table of Contents

Introduction.. 1
Comments on Units and Notation............................ 3

1. A Universal Problem.. 5
   Dealing with Fermi.. 6
   The Drake Equation 9
   "Hey! Over Here!" 11
   Caveat.. 13
   Types of Civilizations 14
   Moving Forward.. 15

2. It's a Matter of Time 23
   Time and Transformation 24
   The Deadly Earth.. 26
   The Deadly Solar System 28
   The Deadly Stars.. 33
   Deadly Novae.. 38
   GRBs and Hypernovae....................................... 44
   The Embrace of Andromeda 47
   Deep Time... 49
   The Doomsday Event 49
   The Long and the Short of It.............................. 50
   Making the Best of It 51
   Dyson Spheres... 52
   Terraforming 53
   Space Structures.. 57
   Thinking Long-Term 58

3. The Sun, Inside and Out 71
   Star Basics 72
   The Dynamical Timescale 73
   Hydrostatic Equilibrium................................... 75
   The Pressure Law.. 76

The Central Temperature 78
Photon Diffusion Time 78
Energy Transport 80
The Surface Temperature of a Star... 82
Stellar Luminosity 83
Energy Generation 83
Nuclear Fusion.. 85
The Mass Luminosity Law 87
A Journey Through the HR Diagram 88
The Journey of the Canonical Sun 90
The Reasons for Gigantism... 95
A Negative Feedback System.. 97
Fundamental Constants 99
The Quantum World of the Electron 101
Collapsing Gas Clouds 102
Why Stars Are Massive 103
A Constraint on Planet Building 105

4. The Price of Doing Nothing... 113
The Habitability Zone... 113
The Ocean on Europa 115
A Brief Aside: Utilizing Europa 117
Moon Life 117
Synchronization and the Moon Effect..................................... 118
The Upper Limit.. 119
A Moving Habitability Zone... 120
The Beginning of the End.. 122
The Fate of Venus and Earth... 123
The Outer Planets.. 125
Orbital Engineering 127
Waving the Flag 130
End Games and Exotic Worlds.. 131
A Moving Imperative.. 134

5. Rejuvenating the Sun 147
The Engineering Options 147
Mixing and Mass Loss 148
Adding to the Pressure 150
The Opacity Effect... 152
The Tools of the Trade... 153
A Homogeneous Star Model 154
Introducing Mass-Loss 155
The Fate of the Ejected Material 156
An Outline Scenario for Rejuvenating the Sun 157
The UV Problem 159

The Extended Solar System Lifetime 160
Mixing It Up 162
Black Hole Mixing 163
A Steady Stellar Diet 165
Brave New Worlds 169
An Alien Beast Within 171
Solar Wrap Control 173
What the Future Holds 175

6. Stars Transformed 181
Revisiting Carter 181
An Exoplanet Review 181
The Case of the Blue Stragglers 184
The Time of Their Lives 189
Under Construction 190
On the Threshold 191

7. Between Now and Then 197
Do We Have a Near-Term Future? 197
What Price the Future? 198
Thinking Long-Term 201
Taking the Next Step 202
Future Earth 205

Epilogue 209

Glossary of Terms 213

Appendix A: A Homogeneous Star Model 217

Appendix B: An Accreting Black Hole Model 221

Index 223

Introduction

This book is about an audacious idea: *asteroengineering*—literally, the physical engineering of a star, especially the star we call our Sun. It is an idea on the grandest of scales. Part science fiction, part science fact, asteroengineering is a response to a very definite and a very real problem, a problem that our distant descendants will one day have to face. It is also a universal problem that will be experienced – at some stage or other – by every extraterrestrial civilization that has or will exist. Indeed, the problem to be addressed resides within the parent stars of each and every life-supporting planetary system within our galaxy. In short, stars puff up to become luminous red giants as they age, and by doing this they vaporize those planets previously situated in the habitability zone where life can otherwise thrive. As their parent star ages and approaches the red giant phase, a civilization has two options open to it: stay at home, or pack up and leave. The latter option would require the hapless civilization to cocoon itself within giant space-ships and then set itself adrift in the uncharted depths of space. If a civilization chooses to stay put, however, then all life will end—unless, that is, something is done about the demise of its parent star.

The idea of star engineering was possibly first discussed in the mid-1980s in the book *Atoms of Silence: An Exploration of Cosmic Evolution* by Hubert Reeves (MIT Press, 1984). The blatant defiance of the idea was inspiring. That we might engineer the Sun – the hub of the solar system and steadfast provider of warmth for life on Earth – now that's an ambitious goal! It is a brazen and challenging notion, and one that sets the mind both searching and reeling. This book provides a preliminary examination of the solar rejuvenation options that may be realized and put into practice by our descendants in the deep future.

1

It is often said that science fiction is just pre-science fact. There are few, if any, science fiction stories about engineering stars to save a planetary system, but this is essentially the aim of asteroengineering. If nothing is done about the future evolution of our Sun, then it will destroy all life on Earth. This fiery destruction of Earth won't happen within our lifetime, however, and we are still many hundreds of millions of years away from potential destruction. Indeed, our distant descendants will have many long years of preparation time before even the first steps towards solar rejuvenation must be taken. Some of our galactic companions – assuming that they exist – will not, however, be as lucky as us. They may, in the here and now, be involved in the very process of altering their parent stars. As we shall argue in Chapters 1 and 6, asteroengineering may, in fact, be in common practice throughout the Milky Way galaxy (and other galaxies), and if so, this can be offered as a potential solution to the famous paradox, posed by physicist Enrico Fermi, which asks why are there no alien life forms in our solar system at the present time. Our suggested answer is that they are not here – that is, in the solar system – because they have had no reason to leave their home worlds. Indeed, it is our contention that advanced galactic civilizations will most likely choose to rejuvenate their parent stars into long-lived, non-giant forming states, than adopt a galactic colonization program.

Before our descendants have to worry about the effects of an aging and more luminous Sun, there are a multitude of terrestrial and astronomical disasters that they will have to guard themselves against first. Not just earthquakes, landslides, tornadoes, and tsunamis; our descendants will have to contend with comet and asteroid impacts, the explosion of nearby supernovae, and the close passages of wayward stars. All of these problems, however, just like the increase in size of the aging Sun, are potentially fixable by strategic planning and, in some cases, direct intervention. Asteroengineering is one of the direct intervention cases. We cannot possibly perform the required engineering at the present time— nor indeed, do we need to. But we can determine what must be done in principle, and that is half the battle. In the meantime, by learning how to tame and nullify the dangers posed to life on Earth by the heavens around us, humanity can begin to acquire the engineering skills that will eventually be needed to save the entire planet and all the life that resides on, in, and around it from the raging gigantism of an aging Sun.

Although you will find technical arguments and a good number of equations written in this book it is hoped and intended that the material presented is accessible to the non-specialist. The mathematics presented is no more advanced than that of a first-year university-level science course, and no detailed knowledge of physics is assumed. The general reader, though, need not worry about the mathematical and technical details too much; if you can't follow the equations then skip to the conclusions, where all should be made clear in words. Chapter 3 is by far the most difficult chapter with respect to its weight of mathematics and physics, but please don't be put off; have a go at trying to follow the arguments. Stars are indeed wonderful objects, whether described in the flowing lyrics of iambic pentameter or precisely described in an intricate web of mathematical detail.

Comments on Units and Notation

Astronomers are notoriously bad for their inconsistent use of physical units. Although we will ostensibly use the *SI* units of meters, kilograms, and seconds, there are times when other (i.e., non-*SI*) units are more conveniently adopted. Distances, for example, will typically be expressed in either astronomical units (AU) or in parsecs (pc), and occasionally as light-years (ly). Accordingly:

$$1 \text{ AU} = 1.496 \text{ x } 10^{11} \text{ m}$$

$$1 \text{pc} = 206, 265 \text{ AU} = 3.261 \text{ ly}$$

We will also use solar units (designated by the symbol $\odot$) where mass, size and energy output per unit time (luminosity) are expressed according to the measured quantities:

$$1 \ M_{\odot} = 1.9891 \text{ x } 10^{30} \text{ kg}$$

$$1 \ R_{\odot} = 6.96265 \text{ x } 10^{7} \text{ m}$$

$$1 \ L_{\odot} = 3.85 \text{ x } 10^{26} \text{ Watts.}$$

Temperatures will be expressed in Kelvins (K), where zero Kelvins (the convention is to say Kevins rather than degrees Kelvin) corresponds to the absolute zero point temperature. In the more commonly used Centigrade (°C) scale, absolute zero falls at -273 °C. Among the mathematical symbols that you will commonly see: '$\sim$,' meaning to order of magnitude, and '$\approx$,' meaning approximately equal to. The symbols '<' and '>' are used to indicate the 'less than' and 'greater than' inequalities. In this manner, for example, a > b means quantity 'a' is greater in magnitude than quantity 'b.'

1. A Universal Problem

Los Alamos Laboratories, New Mexico, 1945. It is lunchtime. Our gaze is directed towards a quiet corner of the otherwise busy refectory hall. A cluster of crop-haired physicists are seated around a large circular table. In twos and threes they huddle together in deep conversation. Snippets of their discourse drift into range. The discussion topics vary wildly: the weather, the ending of the war, a weekend hike, the solution to a particular integral equation, and the many other problems of their work. At some moment, no one is quite sure how or why it happened, one of the physicists, Enrico Fermi,[1] asks a question about extraterrestrial life. "What's that, Fermi?" a voice queries. "I was just thinking" repeated Fermi, "if the galaxy is full of extraterrestrial civilizations, then why are there no extraterrestrial beings on Earth—walking amongst us now?" "Don't be absurd Fermi," one of the group counters. "The galaxy is huge, and it would take longer than the age of the universe to colonize it—surely?"

"Are you so certain?" rejoins Fermi. "Let's do a back of the envelope calculation." There is a flurry of movement among our huddle of physicists. Fermi is famous for his order of magnitude calculations. "Let's assume that there are 10^{10} stars in our galaxy and that each star is on average, oh, say, 6 light years from its nearest neighbor," Fermi begins.[2] "Let's also assume that a spaceship has been developed that uses standard rocket power to propel it at speeds of, oh, let's say, 30 kilometers per second or one-ten-thousandth the speed of light.[3] So, travel time being distance divided by speed gives about 65,000 years to move from any one star to its next nearest companion. A long time by human standards for sure, but small fry compared to the age of the galaxy."

"Not an exactly profound result," someone mumbles, but a series of sharp glances and furrowed brows quells the interrupter.

"Now, as I see it," Fermi continues, ignoring the young upstart, "any reasonable colonization strategy would work like a fission process. One spaceship sets off from the home planet. When it reaches the next nearest star, two new spaceships are made, and these are sent off to the next two nearest stars, where the same doubling process continues. In this way, after 33 generations, every star in the galaxy should have been visited, since 2^{33} is about 10^{10}. So, as I see it, the total time required to visit every star in the galaxy should be of order 33 x 65,000, or 2 million years.[4] Two million years, why compared to the age of the Sun (which is something like 4.5 billion years old) this is a mere nothing. Even on a geological time scale, this is peanuts. A civilization that arose on a planet orbiting a star formed, say, a billion years before our Sun formed, could have easily colonized the entire galaxy. So, I ask again, where are they?"

A hush falls over his audience. Fermi has them in a bind. The younger physicists at the table are racking their brains in the hope of finding a loophole in the argument, hoping to score points with their companions. But no one can come up with a counter answer.

Dealing with Fermi

The lunchtime conversation as just described may never have actually happened, but it is loosely based upon a story recounted by planetary astronomer Carl Sagan. Real or not, however, it is an anecdote that has become encased in legend. Where and when Fermi first raised the topic is not so much the issue; the point is that Fermi's Paradox – as his question has become known – is a real problem that correspondingly requires a solution. A paradox, the dictionary indicates, is a statement that seems contradictory but contains an element of truth. In this manner Fermi's question does require a few points of qualification: it is only a paradox if extraterrestrial beings actually exist and are capable of, and of a mind to, explore and possibly attempt to colonize the galaxy. We shall argue later in this chapter and indeed throughout this book that Fermi's Paradox is answered, in fact, in terms of a no-need-to-colonize principle at work within the galaxy. Indeed, it is the viewpoint of this author that extraterrestrial civilizations will first engineer their parent stars into long-lived states, rather than—or, indeed, instead of—embarking on galactic colonization programs.

This being said, it seems worthwhile at this stage to spend a little time reviewing a few of the basic issues that arise from a consideration of Fermi's question in general.

As of the time of this writing, we do not know if any other life forms exist beyond Planet Earth—intelligent or otherwise. We do, however, have knowledge of one fact, and it is that there are currently no extraterrestrial intelligent beings making their presence clearly known to us on Earth. There is also no clear evidence for extraterrestrial beings ever having visited Earth in the past. This latter point, while true as it stands, should be tempered by recalling the well-known maxim: absence of evidence is not evidence of absence.

With these very limited facts to deal with, however, what might be said in answer to Fermi's question? Well, it is probably fair to say that a very impressive and a very diverse array of solutions have been offered over the years,[5] and they range from the extreme idea that there are absolutely no intelligent life forms, other than us, in the entire galaxy (and universe), to the idea that advanced civilizations are everywhere in the galaxy, but they are being very careful to avoid contact with us. There is also the idea that we happen to live in a very special epoch, which argues that the conditions necessary for a galaxy-colonizing civilization to emerge have not, as of yet, been satisfied. All such explanations are indeed possible solutions to Fermi's Paradox, although some of the arguments seem much more compelling than others.

Perhaps the most incredible and thought-provoking solution to the paradox is that no alien civilizations exist. As Stephen J. Gould once aptly put it, "Perhaps we are only an afterthought, a kind of cosmic accident, just one bauble on the Christmas tree of evolution."[6] The suggestion that we live in some special epoch is one in general to be wary of, no matter how compelling the argument might sound, since it runs against the basic tenant of the so-called Copernican Principle which, under admittedly different circumstances, many astronomers take to be an underlying axiom of their work. This principle, while sounding profound, is essentially a straightforward statement concerning special conditions. Just as Nicolaus Copernicus, in 1543, argued that Earth is not located at the center of the universe but is instead a planet in orbit around the Sun (which he believed was located at the center of the universe), the Copernican Principle in general says that we shouldn't assume that there is anything special about our existence or our location in the galaxy (and the universe). The point behind this principle is that the conditions that have resulted in our existence, while wholly remarkable and exceptionally special in their development, should operate elsewhere in the galaxy, and, consequently, there is no specific reason why other life forms shouldn't have come into existence on planets orbiting other stars within our galaxy (and in other galaxies).

There is a caveat to the Copernican Principle that is worth considering at this stage. Although it makes good scientific sense to avoid special-case explanations for any given observed phenomena, there are circumstances where special conditions must hold true for some specific phenomenon to be observable in the first place. This set of circumstances is often referred to as the Anthropic Principle.[7] Our existence as intelligent observers, for example, is special in that the convoluted set of events leading up to our emergence are not likely to have happened much earlier than they did. That is, the fact that we observe the Solar System to be 4.56 billion years old is in part a necessary reflection of the time required for us to appear. If the time for us to evolve on Earth is T_{us}, then we know $0 < T_{us} < T_{ms}$, where T_{ms} is the main-sequence lifetime of the Sun. We will define what is meant by the main-sequence lifetime, and why it is an upper time limit, more clearly in the next chapter. For the moment, suffice it to say that T_{ms} (Sun) $\approx 10^{10}$ years and that after this time the Sun's luminosity increases so much that Earth's oceans will boil away and all life will perish (see Chapters 3 and 4). The lower bound is clear enough in the sense that we did not evolve instantaneously. Indeed, our immediate ancestors, *homo sapien*, only emerged a few hundred thousand years ago. The so-called Weak Anthropic Principle explains why we observe $T_{us} < T_{ms}$ on the basis that if it wasn't so, then we wouldn't exist to make the observation. This is all rather trivial and self-evident, but the Anthropic Principle has also been used to argue that there are no extraterrestrial civilizations within our galaxy. In particular physicist Brandon Carter has presented the following argument: assuming that the average time T_{av} needed for an intelligent observer to appear on an Earth-like planet is typically much greater than the main-sequence lifetime of a Sun-like star (i.e., $T_{av} > T_{ms}$), but that the longer the evolutionary process proceeds the more likely it is that an intelligent observer is going to appear, then the appearance time T_{IN} for intelligent life should be $T_{IN} \approx T_{ms}$. In other words, intelligent observers are most likely going to appear at the exact same time that their parent star destroys their home planet. According to this line of reasoning the existence of extraterrestrial life forms is, in fact, highly unlikely. Carter suggests that our existence ($T_{us} < T_{ms} < T_{av}$) indicates that our presence is the result of a set of highly improbable evolutionary circumstances, not likely to be repeated anywhere else in the galaxy—ever! So, in some sense, we are special, albeit apparently improbable, after all.

The Drake Equation

When a physicist or mathematician writes down an equation, it is usually because it makes a very definite and precise statement about some particular set of circumstances. When we write down the Drake equation, however, which expresses the possible number of extraterrestrial civilizations N within our galaxy, we ultimately write down an equation that expresses our total ignorance. This statement is not made as an indictment of SETI pioneer Frank Drake,[8] who first discussed the equation now named after him, but a comment upon the fact that we simply don't know, even to order of magnitude, what actual numbers to place in the equation. (This is actually a counter example to Fermi's maxim discussed in Note 1.) There are at least seven terms that can be included in the Drake equation. The first term accounts for the formation rate of stars R^* in the mass range $\sim 0.5\ M_\odot$ to $\sim 1.3\ M_\odot$. We shall explain in Chapter 4 why this particular mass range is important. Having formed a star, there is then a term f_p that accounts for the fraction of those stars that actually have planets. A third term n_L then accounts for the number of planets that reside in the habitability zone (again, discussed in Chapter 4) where life might possibly exist. Three other fractional terms are then introduced: f_l, which accounts for those planets in the habitability zone that actually evolve life; f_I, the fraction of those life forms evolved that actually acquire an advanced 'intelligence,' and f_E, the fraction of those intelligent species that develop advanced technologies (such as radio transmitters producing signals that we might detect). A final term L is introduced to account for the lifetime of the civilization ended by natural or self-destruction. Drake's equation is the product of all these terms:

$$N = R^* \, f_P \, n_L \, f_L \, f_I \, f_E \, L \qquad (1.1)$$

The simplicity of Equation (1.1) is, as mentioned above, deceptive. Surely, all we have to do now is plug in reasonable numbers for all of the terms, and we have our estimate for N. This would be true, of course, if we actually knew with any certainty what values to give for the various terms. The only term that astronomers can place a reasonably good value on is that for R^*, the star formation rate. The second astronomical term f_P is beginning to be constrained through the ever-increasing number of extrasolar planets being discovered. This latter constraint, however, is currently not especially helpful, since only gas and ice-giant Jupiter- to Neptune-like planets have been discovered with any certainty. As of this writing there is no clear idea how many of the presently detected planetary systems might contain terrestrial planets in a habitable zone.[9]

The available observations suggest that the star formation rate R^* has been decreasing ever since the formation of our galaxy (thought to be at least 13 billion years old). This has a number of interesting consequences. The metals (by which astronomers mean all the elements other than hydrogen and helium) that are vitally important for making terrestrial-like planets and ultimately allowing life to come about are all produced through fusion reactions (see Chapter 3) within the cores of massive stars and their end stages as supernovae. Most of the metals were produced, therefore, in a strong initial burst of star formation in the first few 100 million years of the galaxy forming. Consequently it is possible that at least some terrestrial planets may have formed very early on, allowing for the possibility that some extraterrestrial civilizations may have existed for billions of years prior to the formation of our Solar System. This is a thought-provoking possibility indeed, and one that makes the absence of extraterrestrials on Earth – here and now – all the more intriguing, as Fermi noted.

There is an interesting mix of terms in Equation (1.1), and these speak to its hidden complexity. Although the first three terms are essentially astronomical in nature, the terms f_L and f_I are determined according to biological constraints. The last two terms, f_E and L, depend upon the sociology of the intelligent life that has evolved. All we can say currently is that N is at least equal to one—i.e., philosophical issues aside, we exist. As to whether N, after a final tally is made (which begs the question, "how would we know?"), will still be equal to one or if it will be as large as, say, one million is completely unclear at the present time. Either way, even finding out that $N = 2$ would have profound effects upon human society.

"Hey! Over Here!"

A number of researchers have suggested that rather than the actual beings from an advanced extraterrestrial civilization exploring (or colonizing) the galaxy, they might send out self-replicating machines instead. These so-called von Neumann[10] machines are almost mythical beasts, endowed with superior engineering skills and an intelligence that far exceeds those of, say, a mere human being. With these machines we break with the traditional science fiction precedent and submit that just because such mechanisms can be dreamed of does not mean that they will ever be constructed (for galactic exploration, that is). If a civilization wants to actually communicate its existence or colonize the galaxy, then the use of self-replicating machines is an inherently difficult way of trying to do it. Highly advanced robotic and artificially intelligent systems will almost certainly be developed, and they will also probably be used to explore and help colonize a home planetary system, but only under circumstances where profits – commercial and social as well as scientific – can be extracted.

Having argued the above it should be noted, however, that there are actually good reasons for sending spacecraft into inter-stellar space. The difference between spacecraft and Von Neumann machines, however, are that the former are passive and relatively cheap, while the latter are invasive, arguably aggressive, and highly expensive. Neither object actually provides any return of information, but the simple, slow spacecraft method provides an exceptionally efficient means of providing a large amount of information, to any potential finder, in one simple dose. Indeed, when the speed at which any communication proceeds is not of overriding importance, Christopher Rose and Gregory Wright[11] argue that "inscribed matter" messages (Figure. 1.1) are far more

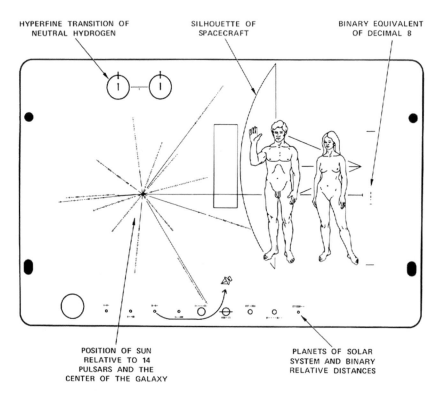

FIGURE 1.1. The plaque carried aboard the *Pioneer 10* and *11* spacecraft, both of which are now traveling into interstellar space. The plaque was designed by Frank Drake in collaboration with Carl and Linda Sagan and carries a wealth of 'inscribed' information. Two human figures are shown to scale by the spacecraft, and a diagram (to the lower right) shows which planet in our solar system the spacecraft came from. The set of 'star' lines towards the center right of the plaque indicate the position of our Sun with respect to nearby pulsars—each pulsar being identified by the binary value of its spin period. The two circles to the upper right of the plaque correspond to the atomic hydrogen molecule. The plaque is about 15cm x 23cm in size. While not actually aimed at any specific star system, *Pioneer 10* is currently heading in the direction of the constellation of Taurus. *Pioneer 11* is heading towards the constellation of Aquila. (Image courtesy of NASA)

cost-effective and efficient than any other mode of communication. Indeed, Rose and Wright comment that "Carefully searching our own planetary backyard may be as likely to reveal evidence of extraterrestrial civilizations as studying distant stars through telescopes."[12]

Caveat

One of the central tenets of the Fermi Paradox is that there is no evidence that extraterrestrial beings or, for that matter, autonomous extraterrestrial spacecraft (such as von Neumann machines) have ever visited our solar system. Although this last statement is true as it stands, there is, of course, the issue of recognition. How can we be sure that all the correct places have been searched when we don't actually know what it is we are looking for? Physicist Stephen Wolfram has recently argued, in fact, that recognizing extraterrestrial intelligence may even be impossible— at least, that is, with current search strategies.[12]

In spite of Wolfram's pessimism, various programs have been initiated in recent years to address the visitation issue. For example, Search for Extraterrestrial Artifacts (SETA) and Search for Extraterrestrial Visitation (SETV) programs have joined the multitude of Search for Extraterrestrial Intelligence (SETI) initiatives already in place. But to date no credible artifacts or visitation data have been found.

The ability to search for and potentially recognize extraterrestrial artifacts, such as spacecraft with inscribed, information-laden messages will, presumably, improve with time. As humans explore more of the Solar System in greater and greater detail, the chances of our finding any embedded artifacts will improve. Since longevity of survival is paramount for any inscribed message platform, Alexey Arkhipov of the Institute for Radio Astronomy in the Ukraine has suggested that the best first-place to look for such objects is our Moon.[13] The Moon, Arkhipov argues, provides shielding from a large fraction of the micro-meteoroid flux, a stable land mass (large impact cratering events aside) that has no atmospheric or biological factors to corrode or disturb equipment.

Arkhipov has also suggested that an "Astroinfect Principle" could be at play within our galaxy.[14] Specifically, Arkhipov notes that the winds associated with stars might eject space debris (literally, the small scraps of paint, fuel pellets, fecal matter, and exotic alloy flecks blasted by meteoroid impacts upon spacecraft in orbit around Earth) into interstellar space, and these alien scraps might just be detectable in, say, the lunar regolith. Ian Morrison writes that "For perhaps 7 billion years there have been enough heavy elements within the interstellar medium for planets to form and intelligent life to arise. If, in this period, one such civilization came into existence every 100,000 years, then 70,000 advanced civilizations might have come and gone. Could one of them have left any evidence of their existence?"[15] One can always argue about the numbers of possible extraterrestrial civilizations, but it is intriguing, and indeed sobering, to think that the past heights of advanced extraterrestrial intelligences within our galaxy might only be betrayed to us by the garbage that they left behind.

Types of Civilizations

Russian astrophysicist Nicolai Kardashev has proposed a three-tiered system of civilization classification. His scheme is based upon how much energy a specific civilization can draw upon.[16] Using Earth and the Sun as the basic measure of energy being consumed and energy potentially available, Kardashev suggests the following designations:

- **Type I** A civilization that can draw upon and consume $\sim 10^{17}$ joules of energy per second. Earth presently corresponds to just such a civilization.
- **Type II** A civilization capable of harnessing and using the entire energy output from its parent star. In the case of our Solar System, this would correspond to the consumption of about 4×10^{30} J/s worth of energy. A civilization capable of building a so-called Dyson sphere (discussed in more detail in the next chapter) about their parent star would correspond to a Type II civilization.

- **Type III** A civilization that can draw upon the energy available on a galactic scale. This corresponds to an energy consumption rate of something like 10^{41} J/s.

Although the actual amount of power available to a civilization will no doubt vary from one stellar system to the next, the basic idea of the scale advancement is clear, with the classification type increasing each time the energy consumption rate jumps from planetary to parent star to multiple star to galactic. It seems that there is no Type III civilization currently in existence within our Milky Way galaxy, and there is currently no evidence to indicate that any Type II civilizations exist (but see Chapter 2).

Moving Forward

In the highly charged game of debating the existence of extraterrestrial civilizations one has eventually to – openly or obliquely – make a statement about how one is going to proceed. We are going to assume, since we have no real proof, that life is probably abundant throughout our galaxy (and other galaxies), and that many advanced civilizations exist with technologies well in advance of our own. We shall also proceed on the basis that no advanced civilization has ever attempted, or perhaps more strongly, has never needed to colonize the galaxy. These working assumptions we will attempt to bolster in the next several chapters. Mostly, however, we will proceed on the basis that it is unlikely that galactic colonization is ever likely to be advanced upon pure economic grounds. Likewise, we will argue that any civilization that survives against self-destruction would have no overriding reason to leave their home planet (or more exactly their planetary system) because of ecosystem collapse. To survive in the long-term requires that a civilization must live in at least partial harmony with its surroundings and have a stable, well-supplied, and well-nourished population. Anything less will inevitably lead to strife and ultimately self-destruction.[17] There are, it seems, fundamental limits to communication and space travel speed that cannot be broken. Specifically, for example, let us assume that space travel will never proceed at speeds anywhere near that of the speed of light.[3] The advanced civilizations that do exist, we suppose, will explore their immediate planetary systems and will carefully utilize all the properties and resources available to them. These civilizations may also build small worlds that orbit their parent star, and they may even build O'Neill and Dyson sphere-like structures (described in the next chapter). In this fashion so-called Kardashev Type II civilizations might eventually be discovered, but we suppose that Kardashev Type III civilizations, utilizing the entire energy of their host galaxy, will not be found.[18]

With all the above being laid out, what is the solution to Fermi's Paradox being developed in this book? In a nutshell, it is this: although the galaxy contains many (or even just a few; it matters very little at this stage) advanced civilizations, they are not among us now because they have never had a need to move away from their home planet (or more precisely, their home planetary system). In addition, for civilizations more ancient than our own, star-engineering has eliminated the imperative towards galactic colonization resulting from the gigantism experienced by non-engineered stars at the end of their main-sequence phase.

We will argue in the following chapters that advanced civilizations will not be forced to seek new planetary systems as a result of their parent stars reaching the end of their main sequence phase. Indeed, advanced societies will invest their superior skills and knowledge into the engineering of their parent star. They will literally control their own destiny and, thus, have no need to seek new havens beyond the reaches of their immediate planetary system. The interstellar exploration that any advanced extraterrestrial civilizations might ultimately undertake will most likely be purely parochial and predominantly scientific fact-gathering missions.

Notes and References

1. Enrico Fermi (1901–54), an Italian-American physicist, is principally known for his pioneering research relating to the modern day understanding of atomic structure. In 1926 he described the statistical law (now called Fermi-Dirac statistics) that governs the behavior of particles subject to the Pauli Exclusion Principle. Fermi was one of the key scientists behind the Manhattan Project leading to

the development of the first atomic bomb. Fermi was also famous amongst his peers for his approach to making 'ballpark estimates' of unknown quantities. Essentially, Fermi argued, that even if you don't know the order of magnitude answer to a question, you can still proceed to estimate the answer by making a series of assumptions about the numbers going into the answer. The law of probability then dictates that while some estimates will be too big, others will be too small and correspondingly the eventual result should be about right. There is a very enjoyable and highly recommended introductory chapter on solving Fermi problems in the book by John Adam, *Mathematics in Nature: Modeling Patterns in the Natural World*. Princeton University Press, Princeton (2003).

2. Observations indicate that the Sun is 8,000 pc from the galactic center about which it orbits with a speed of 230 km/s. From these two quantities the galactic year is evidently some 200 million (Earth) years long. Kepler's third law further tells us that for the Sun's orbital radius and orbital period to have the values that they do, the amount of matter interior to the Sun's orbit must amount to about 10^{11} solar masses. Now, astronomers have also discovered that about 90 percent of this mass is in the form of dark matter that neither emits nor absorbs electromagnetic radiation. The amount of mass in terms of observable matter (i.e., stars) is of order 10^{10} $M_\odot$ therefore. Some of this mass is in the form of the gas and dust of the interstellar medium, some is in the form of hard-to-observe Jupiter-like planets, and some will be in the form of brown and white dwarf stars. Our estimate of 10^{10} stars in the galaxy is probably conservative since most stars have masses of only a few tenths that of the Sun. The average separation of stars is based upon the observed density of about 0.1 stars per cubic parsec volume of space. With this density the typical star separation will be about 2 pc (or ~6 light-years). If we had assumed that there are 10^{11} stars in the galaxy with an average separation of 2 pc then the colonization time increases to about 2.5 million years—a change that has no effect upon Fermi's argument.

3. The speed of light $c = 3$ x 10^8 m/s is a universal constant that defines a limiting speed within our Universe. Our estimate for the speed of an interstellar probe is relatively modest. The New Horizons spacecraft, launched in January of 2006, on its way to study the dwarf planet Pluto will reach a top speed of about 16 km/s. Even at this speed it will take more than nine years to reach Pluto and the Kuiper Belt beyond. If this current upper speed limit of 16 km/s is used in the Fermi calculation (as per Note 2) the galaxy colonization time is

doubled to about four million years – a change that has no significant effect upon the argument being presented.

4. A variety of interstellar colonization models have been proposed and developed over the years and, according to the various input assumptions adopted, a range of colonization times emerge. Typically the galactic colonization time for a single motivated civilization is found to fall between 10^7 and 10^9 years. While these colonization times are 10 to 1,000 times larger than our Fermi estimate, they do not significantly change the argument. The timescales are still such that one would expect aliens to be actively found within our solar system in the here and now.

5. A good – but now a little dated – survey of solutions to Fermi's Paradox is given by Michael Hart, An explanation for the absence of extraterrestrials on Earth. *Quarterly Journal of the Royal Astronomical Society*, **16**, 128–135 (1975).

6. S. J. Gould, *Wonderful Life*, Norton, New York (1989). The famed evolutionary biologist Ernst Mayer (1904–2005) has also pointed out that eyes, for example, have evolved on numerous occasions since life first appeared on Earth, while intelligence (such as ours) has evolved just once. This clearly indicates that the evolution of eyes is highly probable, while the evolution of intelligence, in spite of its great adaptive value, is not. Mayer discusses this point in *Extraterrestrial: Science and Alien Intelligence*, E. Regis Jr. (ed), Cambridge University Press, Cambridge (1985). In contrast to the evolutionary argument, one could also explain our (apparent) uniqueness in terms of religious doctrine.

7. There are a number of forms of the Anthropic principle. The so-called weak form is being presented in this book. By far the most comprehensive guide to the historical development and application of the Anthropic principle is that by John Barrow and Frank Tipler, *The Anthropic Cosmological Principle*, Oxford University Press, Oxford (1986).

8. In 1960 Frank Drake (b. 1930 -) used a radio telescope at the National Radio Astronomy Observatory in Greenbank, Maryland to listen-in to possible radio transmissions from the nearby Sun-like stars Tau Ceti and Epsilon Eridani. Nicknamed Project Ozma, this was the first dedicated radio survey in the search for extraterrestrial intelligence (SETI). Drake developed his famous equation in 1961 on the number of possible extraterrestrial civilizations. The purpose of the equation was (and still is) to focus attention towards the crucial questions that might determine the chances of SETI's success.

9. It is entirely possible that, in principle, life could exist within the upper atmospheres of Jupiter-like planets having orbits that place them close-in to their parent star. Life could also, in principle, exist on a moon in orbit around a Jovian planet. The moon Europa that orbits Jupiter in our solar system, for example, has a global ocean that may harbor life (see Chapter 4). Enceledus, in orbit around Saturn, is another candidate moon that may potentially support life given that subsurface liquid water is thought to be an important component in shaping a number of its observed surface features. The potential number of Earth-like planets within our galaxy is not easily estimated, but recent observational results suggest that they may, in fact, be very plentiful.

10. John von Neumann (1903–1957) was the Hungarian-American mathematician responsible for the development of game theory and the mathematical foundations of quantum mechanics. He was a pioneer of computer science and was a key player in the development of the theory of electronic computation. Barrow and Tipler (Note 7) have especially argued that we (that is, humanity on planet Earth) must be unique in the galaxy since no Von Neumann-like machines have ever visited the solar system.

11. Christopher Rose and Gregory Wright, Inscribed matter as an energy-efficient means of communication with an extraterrestrial civilization. *Nature*, **431**, 47–49 (2004). Commenting on the paper by Rose and Wright, Woodruff Sullivan III [Message in a bottle, *Nature*, **431**, 27–28 (2004)] notes that it has been estimated that all of the written and electronic information that now exists on Earth constitutes about 10^{19} bits of information. If one used scanning tunneling microscopy, as suggested by Rose and Wright, to encode this information in nanometer squares of xenon atoms placed on a nickel substrate, then the entire written knowledge of humanity could be inscribed within 1 gram of material. Certainly this methodology encodes a fantastic amount of information in a very small package, but as Sullivan argues, "We do not know if such a package, even if efficiently sent, would ever be recognized and opened." The point is, of course, that there are times when arguments based upon economics and efficiency, which are little more than the 'spoiled children' of guesswork and ideology anyway, are not enough. Sometimes, the inefficient and more expensive approach will return better dividends.

12. Wolfram's views have been nicely described by Marcus Chown [The alien within your computer, *Astronomy Now*, **20** (7), 32–35, July (2006)]. With respect to electromagnetic communications, Wolfram argues that advanced civilizations will use methods that are far more

complicated and much less structured than the forms we are accustomed to using. In this sense, Wolfram suggests, SETI radio surveys that only 'search' for extraneous periodic signals are probably doomed to failure at the very outset.

13. A. V. Arkhipov, in *Progress in the Search for Extraterrestrial Life*. G. S. Shostak (ed.). ASP Conference Series, **74**, 259–267 (1995). Arkhipov, in fact, argues in his article that, "Landing on the moon would be for ET visitors a necessity rather than a convenience."

14. A. V. Arkhipov, New arguments for panspermia. *The Observatory*, **116**, 396-397 (1996). Arkhipov argues that microorganism-contaminated space debris can potentially form a large, – of order 1-pc in size – non-sterile zone (NSZ) around a star supporting an intelligent, space-exploring civilization. The passage of another star (and its accompanying) planets through a NSZ could result in the planets being infected by space-borne microbes. Arkhipov argues that something like 200,000 stars capable of supporting planetary systems will have passed within 1.5 pc of the Sun since its formation 4.5 billion years ago. We will pick up the theme of stellar encounters in the next chapter.

15. I. Morrison, SETI in the new millennium. *Astronomy and Geophysics*, **47**, 4.12–4.16 (2006). Morrison points out in this review article that, to date, only a very small fraction of the radio spectrum and galaxy has been studied at radio frequencies. The next major step forward for radio SETI is likely to be the completion of the Square Kilometre Array (SKA), a project currently under design study (http://skatelescope.org/).

16. N. Kardashev, Transmission of information by extraterrestrial civilizations, *Soviet Astronomy-AJ*, **8** (2), 217 – 221 (1964).

17. Here I am betraying a somewhat humanistic philosophy which may not be a required dictate for of all civilizations. It is not inconceivable that a civilization (either terrestrial or extraterrestrial) might maintain a rigidly enforced birth control and euthanasia program to ensure its long-term survival. This process is (arguably) fair; all individuals get some allotted time to enjoy a full and happy life, but no individual is allowed to overexploit the resources needed to support the current and future generations. It is possible that highly aggressive and secular societies – such as our own could be characterized – will never achieve long-term survival. To face an extended future or to journey to nearby star systems may require a more non-materialistic, non-commercial, and inherently spiritual outlook. Some of these issues are nicely explored in Mary Doria Russell's fictional book *The Sparrow* [Black Swan, 1998]. Secular societies

certainly achieve many great triumphs, but they never live up to their full potential. We will pick up on these themes again in Chapter 7.

18. James Annis, an astrophysicist at Fermilab in the United States, has conducted a study of the brightness characteristics of several hundred elliptical and spiral galaxies in search of potential Type III civilizations [Placing a limit on star-fed Kardashev type III civilizations, *Journal of the British Interplanetary Society*, **52**, 33–36 (1999)], but no candidate civilizations were revealed. Indeed, Annis goes further and uses the available galactic survey data to estimate that the time required to produce a Type III civilization cannot be less than 300 billion years. Since the Universe is only 13 to 14 billion years old, it seems clear that Type III civilizations simply can't exist within our present universe.

2. It's a Matter of Time

The Sun is the lifeblood of humanity. It warms us and gives us light. Although our everyday existence depends upon the Sun, the evolution of life on Earth was only possible because the Sun and the Solar System are partly composed of the embers cast into space by a long line of stellar ancestors. Indeed, the intricate and multitudinous steps that have paved the way to our existence started as soon as the Milky Way galaxy formed some 12 to 13 billion years ago. Step by tiny step our journey has been traced out. We are both ancient and modern, with our bodies being composed of archaic matter forged in the unimaginably intense fires of the primordial moment of creation (the so-called Big Bang) and from matter transformed and fused in the dense and blistering hot cores of massive stars.

The calcium that is contained within the bones that animate our all-too-temporary bodies was formed in the violence of star explosions. As pioneer astrophysicist Sir Arthur Eddington once so aptly put it, "We are the journey work of the stars." Every atom in the universe, other than those atoms of hydrogen and helium,[1] was made by the Sun's distant ancestors, generated piece by piece from the time of the first stars onward. Earth and the life that teems on, above, and through it; the other planets that orbit the Sun; their attendant satellites, and the swarm of comets that surrounds the Solar System, are all made of material produced by long-dead stars.

The Sun provides Earth with energy. If it were not for the Sun, we would not be here—it's as simple as that. It is also because of the Sun, however, that we may not be here forever. Deep within the Sun there is a demon, a lurking monster that will eventually see release and, if nothing is done to tame it, it will destroy all life on Earth. We are in the midst of a horror movie. The twist, however, is that we know the cause from the very outset. Indeed, the only part of the movie that we don't know is exactly when the horror will be perpetrated. We can be fairly sure the demon won't be released in our lifetime; even our descendants a million generations from now will not necessarily be its victims.

Although we live in the midst of this movie, we will never know the final outcome of the saga. It is our distant descendants (perhaps 50 million generations from now) that will face the demon in the Sun—unless, that is, they do something about it, and for the present we can only guess at how well they might fare. Not only will we not know the outcome of this story, there is the open-ended possibility that no destruction will actually be perpetuated. Perhaps our distant descendants will actually survive the onslaught.

Time and Transformation

What is this demon that lies in the Sun? It is, in fact, the very two-headed demon that led to our existence in the first place: time and transformation. The Sun is converting a staggering (that is, by human standards) 4 billion kilograms of matter (actually hydrogen; see below) into energy every second of the day, day after day after day. Since it formed some 4.56 billion years ago, the Sun has converted about 6×10^{26} kg of matter into energy. That's equivalent to the mass of the planet Saturn and five Earths. It is a ceaseless process; the Sun is hungry, and it feeds upon itself. And yet, the total amount of matter already transformed into energy is a mere 0.03 percent of the Sun's current mass—a minuscule diet and one that it can easily afford to consume.

In order for the Sun to replenish the energy radiated into space at its surface, and for that matter for the Sun to remain inflated, it has tapped into a seemingly inexhaustible energy source—the hydrogen out of which it is predominantly made. The energy is derived via the conversion of hydrogen into helium through the proton-proton chain (described in Chapter 3) of fusion reactions—a beautiful sequence of chance encounters, interactions, and explosive transformations, described once again in prosaic form by Sir Arthur Eddington as "a jolly crockery-smashing turn of a music-hall."[2] Out of the chaos deep within the Sun's core, energy is generated, the Sun remains hot and stable, and in consequence we can live on a gently warmed Earth. Not too hot, not too cold, we occupy the Goldilocks planet of the Solar System. But the demon lurks and grows within the Sun.

The Sun contains within its massive girth something like 1.4×10^{30} kg of hydrogen (and about 0.589×10^{30} kg of material other than hydrogen)—a seemingly endless supply of fuel, but for an ever energy hungry Sun it is not enough. It can never be enough. One day the hydrogen within its core will all be gone, transformed into energy and an 'ash' of helium. This is deep time, some 5 billion years from now, when the demon will be born: a roaring cuckoo-child that will grow into a bloated red giant.

The time that it takes a star to use up the hydrogen supply in its core is called the *main-sequence lifetime*, or T_{MS}. The details relating to stellar structure, internal energy generation, and stellar evolution will be presented in detail in the next chapter. For the moment let us simply take it as a fact that a fraction q of a star's total mass M^* is available in the form of hydrogen to produce energy. The main-sequence lifetime can now be calculated according to the total amount of energy available to the star divided by the rate at which that energy supply is used up. The total fusion energy available to a star can be determined from an experimental result first established by chemist Francis Aston in 1919.

Specifically, Aston, who was working at the Cavendish Laboratory in Cambridge, found that the mass of four protons was smaller than the mass of a helium atom nucleus (composed of two protons and two neutrons) by 0.007 percent of the mass of the protons. From this observation, Eddington realized in 1920 that if nature could coax (somehow) four hydrogen (H) atoms into forming a helium (He) atom, then energy could be made available to a star. Symbolically, 4H $\Rightarrow$ He + E, where E is the energy liberated per conversion. The energy term, Eddington reasoned, would be related to the mass difference found by Aston—that is, using Einstein's famous equation linking mass and energy, $E = \Delta m\, c^2$, where c is the speed of light and $\Delta m = 0.007\, M_H$, and where M_H is the mass of the hydrogen atom. In this fashion, the total amount of energy available to a star is $E_{total} = 0.007\, c^2\, q\, M^*\, X$, where X is the mass fraction of a star that is hydrogen ($X = 0.75$, for our purposes). Now, the rate at which a star consumes its energy supply is given by its luminosity L^*, which accounts for the electromagnetic energy radiated into space per unit time at its surface. Accordingly, we have:

$$T_{MS} \approx E_{total}/L^* = 0.007c^2qM^*X, /L^* \qquad (2.1)$$

For the Sun, detailed numerical models indicate that $q \approx 0.13$ and consequently $T_{MS}(SUN) \approx 3 \times 10^{17}$ seconds $\approx 10^{10}$ years. On this basis the Sun is middle-aged with respect to its main-sequence lifetime, and has some 5 to 6 billion years of hydrogen fuel supply left.

Time and solar transformation—these are humanity's deep-time enemies. Time we cannot control, but the transformation and mixing of matter in the Sun we might just be able to influence, and this proactive engineering option – literally the rejuvenation of the Sun – is perhaps the last best hope for providing our distant offspring with a long-term future, and our best hope for saving Planet Earth from total destruction.

The Deadly Earth

Life is a continuous struggle against the elements and, apparently, there are many more ways of dying than there are of staying alive. It is a narrow path that we have to tread to literally keep life and limb together. Over the timescales of concern in this book, however, it is not the survival of any one individual that is especially important. Rather, it is the long-term survival of humanity and Earth that matters, along with all the incredible diverse species of animals that live upon and in it.

The scale of any disaster can be measured in numerous ways— some more impersonal than others. The scale might be measured in lives lost, or in financial infrastructure undermined. It can be measured in irreplaceable historical artifacts destroyed or future growth delayed. But in a typical human lifetime (a measure of, say, 70 years) the sum total of natural disasters that might be witnessed – be they hurricanes, tsunamis, earthquakes, or volcanic eruptions – will be small fry compared to the disasters caused

FIGURE 2.1. A satellite image of Hurricane Katrina. This one event devastated the city of New Orleans in August 2005. It caused the untimely death of 1,836 people and produced an estimated $81.2 billion in damage and destruction. (Image courtesy of NASA).

by extraterrestrial influences that operate on timescales of many thousands of years and longer. Indeed, on the timescale of tens of thousands of years countless numbers of hurricanes (Figure 2.1) will have whipped across the face of Earth, buzzing over the landscape, in accordance with our accelerated time frame, like a carefree fly on a hot summer's day. Whole segments of continents will have been shaken and shifted by earthquakes, or become covered in ash and blistering lava by volcanic eruptions. But all of these disasters combined will melt into insignificance on the day when a 2-km-diameter asteroid ploughs into Earth's atmosphere. And here the correct adjective is when, not if. This event will happen—it's a certainty, unless humanity does something about it.

The Deadly Solar System

The main belt asteroid region, located between the planets Mars and Jupiter, is a reservoir for planetesimals that failed to be accreted into planets. They are primordial objects formed at the very beginning of Solar System history, and they are the scourge of the inner Solar System. Countless collisions, grazing sideswipes, and small-particle strafing have ground the initial population of asteroids into a vast swarm of blasted and fragmented shards. There are literally hundreds of thousands of asteroids now orbiting our Sun. The meteorites that occasionally career through Earth's atmosphere, in a blaze of light and concussion of sonic booms, are asteroid fragments, and it is by collecting these small fragments that we have learned about the processes by which our Solar System formed.

The chances of an individual being struck by a meteorite are (luckily) very slim. Certainly, a few people have been bruised by meteorites throughout recorded history, but as far as reliable sources go, no one has ever been killed by a meteorite strike.[3] The main belt asteroids are not a direct threat to Earth. The simple reason for this is because they move along circular orbits located between Mars and Jupiter. There is, however, a small subset of asteroids (admittedly derived from the main belt region) that do, however, pose a collisional threat to Earth. These are the near-Earth asteroids (NEAs). The NEAs have more elliptical orbits than their main belt relatives, and this property can bring them into the region of Earth's orbit[4] where collisions (and close encounters) will eventually occur.

The most recent impact of a 10-m-sized asteroid with Earth occurred on September 3, 2004. Military reconnaissance space-craft recorded this particular event, and the material that survived atmospheric passage (fortunately) ploughed into the Antarctic ice shelf. Unfortunately (for science), the fall location of this particular impact means that no ground investigation took place, and the

Figure 2.2. The Tunguska fireball devastation. At 07:17 local time on June 30, 1908, the sky over central Siberia filled with the light of a massive, detonating fireball. In an instant thousands of trees were felled and scorched by a searing blast wave generated by the catastrophic breakup of the meteoroid just 10 km above the ground. It is still a matter of contention between planetary scientists whether the Tunguska object was a comet (predominantly composed of water ice) or an asteroid (predominantly stony in composition). The photograph reproduced here was taken during the 1927 expedition led by Leonid Kulik to study the impact area.

crater and possible fragments are now lost. Historically speaking, the best-studied large impact event occurred in Siberia on June 30, 1908. This – the so-called Tunguska event[5] devastated some 2,000 square kilometers of forest and produced a series of long-wavelength acoustic waves that propagated around the world multiple times (Figure 2.2). The last major event involving a 10- to 15-km-sized impactor occurred some 65 million years ago. It has been argued that it was as a result of this particular impact that the dinosaurs (and indeed, most of the world's large land animals) became extinct (Figure 2.3), although many researchers also argue that the impact was more of a *coup de grace* than the principal agent behind the extinction. Many other abrupt extinction events are recorded in the fossil record and only a few of these have been linked to large impact structures.[6]

The extensive and, indeed, ongoing collisional alteration of the main-belt asteroid population dictates that there are now many more small asteroids than large ones. Earth impact probability increases, therefore, with decreasing asteroid size (or mass). So, for example, Earth experiences a collision with a 50-m-diameter

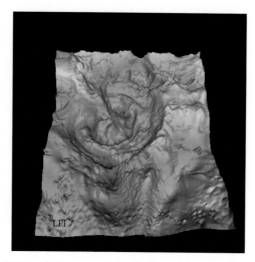

FIGURE 2.3. The Chicxulub impact crater located on the Yucatan peninsula, Mexico. The actual crater is no longer a distinctive surface structure, and the image shown here is a reconstruction of the crater through surface magnetic and gravitational anomaly measurements. The crater is 170 km across and is dated to the time of the great dinosaur extinction 65 million years ago. (Image courtesy of V. L. Sharpton, LPI).

asteroid once every few hundred years; an encounter with a 1-km-sized asteroid occurs every few hundred thousand years, and an impact from a 10- to 15-km-sized asteroid takes place every few tens of millions of years (Figure 2.4). An approximate formula

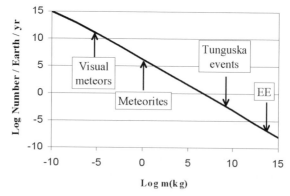

FIGURE 2.4. The total number of objects $N(M)$ with masses larger than mass M, falling to Earth per year. Extinction event (EE) impacts occur on timescales of several tens of millions of years, whereas Tunguska-like events occur on timescales of a few hundred years.

for the time interval between impacts from asteroids of diameter D-km is: T_{NEA} $(yrs) = 10^{[5+0.8ln(D-km)]}$.

Dealing with potential Earth-impacting comets and asteroids will presumably be one of the first priorities of future space engineers. The nature of the danger posed by such objects, however, must first be understood. Indeed, we still need to know which objects might impact Earth. To this end a small number of astronomers, working from observatories dotted around the world, are scanning the sky in the search for rogue asteroids, annotating their numbers and determining their orbits. The orbits and near-term evolution of all asteroids, both main-belt and Earth crossing, with sizes larger than a few kilometers will soon be known, and in principle the times of potential future impacts from these objects will be predictable. Our present knowledge of the asteroid population with sizes less than about 1 km is limited, and the lead-time we might currently expect before the next Tunguska-like impact is perhaps a few seconds. In other words, the first we will know about the event is when it has happened. Although this is far from comforting news, within the next half-century (perhaps) most of the potential Tunguska-like impactors, with diameters of just a few tens of meters, will also be catalogued.

In 1999 MIT researcher Richard Binzel developed a 10-point scheme to measure the impact threat posed by an asteroid. Based upon the estimated size, potential impact velocity and impact probability, the so-called Torino scale developed by Binzel assigns a color and a number to any Earth-approaching asteroid. The Torino scheme is described in Table 2.1. At the time of this writing only one object – an asteroid called 2004 VD17 – has a Torino scale designation greater than white 0. This particular asteroid is believed to be some 600 m across, with a mass of about 2.5 x 10^{11} kg. Having an estimated impact velocity of 21 km/s this asteroid could impart some 14,000 megatons of TNT-equivalent energy upon impacting Earth. The Torino scale designation for 2004 VD17 is Green 1, and while an Earth impact is highly unlikely, a relatively close approach on May 4, 2102, is predicted. It is highly likely that follow-up observations of this particular asteroid, during the next several months and years, will see it downgraded to a Torino scale white 0 object.

Table 2.1. The Torino impact scale. Further details can be found at http://neo.jpl.nasa.gov/torino_scale.html.

Color	Number	Comments
White	0	No direct impact threat. Or a small object (size less than 50 m) that might possibly produce a meteorite shower.
Green	1	Close Earth passage, but very low probability of impact.
Yellow	2	An object warranting further attention, but no great impact concern.
Yellow	3	An object with a > 1% chance of hitting Earth. Impact would produce local destruction. Object warrants further study.
Yellow	4	An object with a > 1% chance of hitting Earth. Impact would produce regional destruction. Object warrants further study.
Orange	5	Close encounter posing a serious threat of impact resulting in regional destruction. Impact lead-time less than 10 years.
Orange	6	Threat of a global catastrophe. Impact lead-time less than 30 years.
Orange	7	Very close encounter by a large object that might hit Earth during the next 100 years.
Red	8	A certain impact, resulting in localized devastation.
Red	9	A certain impact, resulting in regional devastation.
Red	10	A certain collision, resulting in catastrophic global devastation.

Knowing when an impact might occur, however, is only part of the story; what to do about the impending impact is something else altogether. A number of proposals concerning asteroid defense tactics have been put forward in recent years, and they range from the "blast 'em to pieces" type of proposal to the "change their orbit slowly and gently" type. Of course, the action that might be taken is dictated by the size of the asteroid and the potential lead-time before Earth impact is going to occur. If the asteroid is a small one (perhaps a few hundreds of meters across; see Figure 2.5) and the lead-time to an impact is very short (a few years) then blasting it with nuclear weapons is probably the best last-choice action.[7] If the asteroid is large (in excess of, say, 500 m or so) and the lead time is many tens to hundreds of years, then attempts to just slightly alter the asteroid's orbit (for example, by attaching a set of large space sails) would seem to make sense.

FIGURE 2.5. Asteroid (25143) Itokawa is some 0.535 by 0.294 by 0.209 km in size and weighs in at about 3.5 x 10¹⁰ kg. Visited by the Japanese *Hayabusa* spacecraft in late 2005, it is presently believed that the complex structures observed on its surface are the result of a collisional breakup followed by a reassembly phase. That many (possibly most) asteroids are loosely structured rubble piles has important consequences for any potential collisional avoidance strategies. (Image courtesy of JAXA).

The potential impact threat posed by near-Earth asteroids and methods by which such impacts might be avoided are relatively well understood (at least in principle) at the present time. Indeed, the stage is set (literally ready and waiting) for the first attempts to alter the orbit of a 'safe' non-impacting asteroid.[8] The first (terrestrial) asteroid impact avoidance engineers have already been born.

The Deadly Stars

The Sun is situated at a distance of 8,000 pc from the galactic center. It travels through space at a speed of 230 km/s and completes one galactic orbit every 200 million years. On its journey around the galactic center, the Sun is accompanied by an ever-moving host of stellar companions. The nearest star to the Sun is currently the faint dwarf star Proxima Centauri. Located just under 1.3 pc away, however, our nearest stellar neighbor is not actually visible to the naked eye. Within the volume encompassed by a sphere of radius 5 pc centered on the Sun, there are 49 star systems. Indeed, the number of stars per unit volume of space in the solar neighborhood is determined to be $\rho^\star \approx 0.1$ stars/pc^3. This is a useful number, since it enables us to determine the typical time interval (T_{CE}) for close encounters between the Sun and another star.

To determine T_{CE} we need to first set up an impact cross section area: $\sigma = \pi\, d^2$, where d is the encounter distance as measured from the Sun. Given that the Sun has a relative velocity V^* (km/s) with respect to its nearest neighbors, then in time T the volume swept out by the Sun will be $V_\odot = \sigma\, V^*\, T$. In the volume of space swept out, however, there will be $N^* = V_\odot\, \rho^*$ actual stars. Hence, in time T, the number of stars encountered by the Sun within distance d is $N^* = \rho^*\, \sigma\, V^*\, T$. The time interval between individual encounters is, therefore, about $T_{CE} = 1 \,/\, \rho^*\, \sigma\, V^*$.

The question now is, what do we mean by a close encounter? As the approach distance d is increased so the cross sectional area σ increases and T_{CE} correspondingly decreases, so distant encounters become common. The time interval between stars being as close to the Sun as Proxima Centauri ($d \approx 1.3$ pc) is now 74,000 years, given a typical relative velocity of $V^* = 25$ km/s. What about an even closer encounter, say at a distance of 5,000 AU, well inside of the Solar System's cometary reservoir, the Oort Cloud?[9] In this case $T_{CE} \approx 200$ million years. How about an approach as close as 5 AU? At this encounter distance the Solar System will almost certainly be destroyed (depending on the mass of the interloping star), with the planets being scattered out of their current orbits and most probably ejected to a cold, dark death in interstellar space. For this situation, $T_{CE} = 2 \times 10^{14}$ years, which is well in excess of the present age of the galaxy ($T_{gal} \approx 13 \times 10^9$ years). Within the main-sequence lifetime of the Sun [T_{MS}(Sun) $\approx 10^{10}$ years], no star is likely to pass closer than $d \approx 730$ AU.

Although a direct collision – or even a very close encounter – between the Sun and another star is a highly unlikely event, encounters at the distance of the Oort Cloud and of a few thousand astronomical units will happen reasonably often, and these encounters can still be deadly to life on Earth. Cometary nuclei in the outer part of the Oort Cloud will have their orbits gravitationally perturbed by passing nearby stars. The smaller the closest approach distance, the more massive and the slower the speed of the perturbing star, the greater the number of cometary nuclei that will have their orbits changed. Many of the cometary nuclei will be ejected from the Solar System, doomed thereafter to

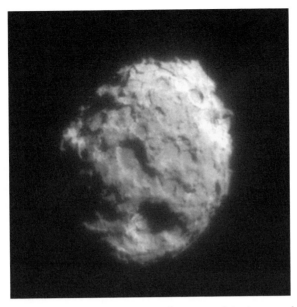

FIGURE 2.6. Close-up of a cometary nucleus. In this image the surface of periodic comet Wild 2 is revealed by a camera carried aboard the *Stardust* spacecraft (http://stardust.jpl.nasa.gov/home/index.html). The nucleus is about 5 km across. (Image courtesy of NASA)

rove the depths of interstellar space. Other cometary nuclei will have their orbits perturbed in such a fashion that they will swing into the inner Solar System, and while on their journey around the Sun they may potentially hit one of the planets (Figure 2.6).

Although Proxima Centauri is currently the closest star to the Sun, we can ask which stars in the solar neighborhood are going to make even closer approaches to the Sun in the future. This question, in fact, was recently answered by Joan Garcia-Sanchez of the University of Barcelona and co-workers.[10] Using data gathered by the *Hipparcos* astrometric satellite, the team found a total of 87 stars that will pass within 5 pc of the Sun during the next 10 million years. Of these, the star Gliese 710 will make the closest approach of all, skimming the edge of the Oort Cloud at a distance of about 0.3 pc (or 70,000 AU) from the Sun (Figure 2.7). The closest approach will occur some 1.4 million years from now. Estimates of the effect of Glies 710 on the Oort Cloud suggest that around 2 million long-period comets will be perturbed into potential Earth-crossing orbits over a time interval lasting perhaps 5 million years.

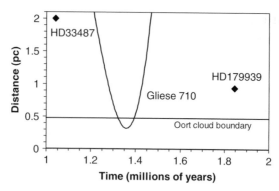

FIGURE 2.7. Closest approach distances of some selected nearby stars. The curved solid line shows the distance variation with time of Gliese 710. The stars HD 33487 and HD 158576 will pass as close as 2 and 0.9 pc respectively to the Solar System. After Gliese 710, the star HD 158576 makes the second closest approach to the Solar System during the next 5 million years.

As this is a chapter concerned with timescales, we can ask the following: What is the likely time interval T_{LPC} between long-period comet hits on Earth? The target area of Earth (ignoring gravitational focusing) is $\sigma_E = \pi R_E^2$, where $R_E = 6371$ km is Earth's radius. The probability that a single Earth-orbit crossing comet might actually strike Earth is, therefore, $P_{hit} = 2\sigma_E/(4\pi D_E^2)$, where $D_E = 1$ AU $= 149.6$ million km is Earth's orbital radius. The factor of 2 accounts for the fact that the cometary orbit cuts through the sphere of radius 1 AU twice, and it is assumed that long-period comets can approach the Sun from any direction (an assumption that is not strictly true for cometary showers). If the number of comets heading into the inner Solar System is f_{com} comets per year, then the time interval between possible Earth impacts will be of order $T_{LPC} = 1/[(R_E/D_E)^2 f_{com}/2] \approx 10^9$ (yr)$/f_{com}$.

The typical flux of long-period comets approaching to within 1 AU of the Sun is estimated to be $f_{com} \sim 100$ per year. (This may rise to ~200 per year during a cometary shower such as that induced by Gliese 710.) These numbers indicate that long-period comet impacts upon Earth are going to be rare, even under cometary shower conditions, with $T_{LPC} \sim 5 - 10 \times 10^6$ years.

The situation with long-period comet impacts is actually more complicated than indicated above. Oscillations of the Solar System above and below the galactic plane (with a period of about 35 million years), gravitational tides resulting from galactic spiral arm encounters, and the close passage of giant molecular clouds can also perturb the orbits of cometary nuclei in the Oort Cloud, causing cometary showers. Mass extinction events deduced from the fossil record (Figure 2.8) and the estimates of terrestrial crater ages suggest that cometary impacts occur in discrete bursts, lasting for perhaps a few million years, separated by intervals corresponding to 25 to 30 million years.[11]

Long-period comet impacts may well be rare, but they pose a tremendous threat to life on Earth. Their large, kilometric size and high encounter velocities indicate that they can deliver devastatingly large amounts of impact energy. Not only this, the lead time between the detection of a long-period comet and its possible impact with Earth might be just a few months. This is a very different situation to the NEA threat discussed earlier, and there is no obvious engineering option that can easily save Earth from a

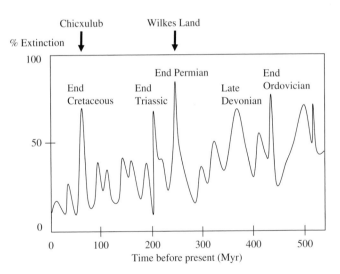

FIGURE 2.8. Schematic diagram of marine species extinctions as recorded in the fossil record over the last 550 million years. The so-called 'big five' events are labeled. The dinosaurs and the Chicxulub impact crater are dated to the end-Cretacious extinction some 65 million years ago. The greatest mass extinction occurred at the end of the Permian, about 250 million years ago. It has been suggested that the end-Permian extinction coincided with the formation of the 480-km-diameter Wilkes Land impact crater recently found in eastern Antarctica. The ripple effect, with a period of about 25 to 30 million years, is clearly visible in the extinction curve.

FIGURE 2.9. On July 4, 2005, the *Deep Impact* mission produced a new impact crater on the surface of Comet Tempel 1. The comet is some 14 km x 5 km x 5 km in size and predominantly composed of water ice. The impact was produced through the direct hit of a 1-m-long, 1-m-diameter copper cylinder striking the icy surface at about 10 km/s. The impact crater produced is thought to be about 100 m across. (Image courtesy of NASA)

direct hit by a long-period comet. Brute explosive force is perhaps one solution (Figure 2.9), but the multiple fragments produced by such actions may well exacerbate rather than fix the problem. Indeed, one of the most pressing issues related to the long-term survival of humanity is how to quickly (on a time scale of months) alter, albeit very slightly, the orbital track of a directly impacting long-period comet without breaking it apart.

Deadly Novae

The end phases of some stars are very violent. They literally blow themselves to pieces (Figure 2.10), spraying processed nuclear material into interstellar space at immense speeds and irradiating vast volumes of space with lethal X- and γ-ray radiation.[12] There

FIGURE 2.10. The Crab Nebula, the icon of a supernova remnant. The 'New Star' associated with the production of the Crab Nebula was recorded in numerous Chinese and Korean chronicles, and was first seen in July 1054. The object was visible for many months, and was reportedly visible in broad daylight. At a distance of 2,000 pc, the maximum brightness of the nova would have been about magnitude -6, brighter than the planet Venus but not as bright as a half-illuminated Moon. (Hubble Space Telescope image, courtesy of NASA)

is some debate concerning the minimum distance beyond which Earth's atmosphere might be safe from a nearby supernova event, but it is typically thought to be between 50 and 100 pc.

An approximate timescale for nova irradiation can be made as follows. Here we will work in two dimensions only and consider the galactic disc to be a circular band with an inner radius of 1 Kpc and an outer radius of 15 Kpc. The area of this galactic disk is then $A_{gal} \approx 7 \times 10^8$ pc^2. If the critical distance from the Sun for a nova detonation is taken to be R_{crt}, then on average the number of nova N_{NVA} that must occur before one is at least within a distance R_{crt} from the Sun is $N_{NVA} = A_{gal}/(\pi R_{crt}^2)$. If we take $R_{crt} = 60$ pc, then

$N_{NVA} \approx 6,000$. Taking a disk nova rate of $S_{NVA} = 0.04$ per year, then the typical time interval T_{NVA} between critical nova explosions is going to be $T_{NVA} = [A_{gal}/(\pi R_{crt}^2)]/S_{NVA} \approx 1.5 \times 10^6$ years.

The next closest Type I supernova-producing system[13] to us may have already been identified, and it is the binary star known as IK Pegasus (also designated HR 1820). In this particular case one of the stars in the binary has already formed into a white dwarf, while its companion is presently in a subgiant phase. The two stars are sufficiently close (just 42 $R_\odot$ apart, in fact) that the subgiant star is actively adding mass to the white dwarf star. This latter accretion is the critical point. In 1930 Nobel Prize-winning physicist and mathematician Subrahmanyan Chandrasekhar showed that there is a maximum mass limit for a white dwarf to remain stable. The critical mass – the Chandrasekhar limit – beyond which a white dwarf will catastrophically collapse is $M_{CLM} \approx 1.4$ $M_\odot$. Once $M_{WD} > M_{CLM}$ then gravitational collapse will ensue, and a Type I supernova will result (Figure 2.11).

It is estimated that the white dwarf component in IK Pegasus has a mass of about 1.2 $M_\odot$, and this means that once it has accreted an additional 0.2 $M_\odot$ of material from its companion a dramatic collapse will occur. If the accretion rate is as high as 10^{-6} solar masses of material per year, then IK Pegasus will undergo supernova disruption in about 200,000 years from now, at a distance of some 44 pc—well inside the 'danger zone' for perturbing Earth's atmosphere (Figure 2.12). For an accretion rate of 10^{-7} solar masses per year, supernova disruption will occur about 2 million years from now, when the system is at a distance of 43 pc from us. The rate at which the white dwarf star in IK Pegasus accretes matter from its companion must be less than 5×10^{-8} solar masses of material per year if supernova disruption is to take place at a distance when the system is beyond the safe distance boundary of 60 pc. At the present time astronomers do not know what the accretion rate for the white dwarf in the IK Pegasus system is.

The speed with which material is blasted into space by a supernova is typically several thousands of kilometers per second. Clearly if a planet is within a few tens of parsecs of a supernova explosion and chances to encounter such a disruptive blast wave, then its atmosphere would be severally denuded.

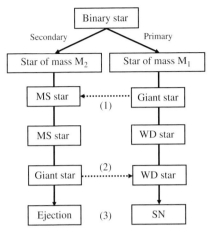

FIGURE 2.11. Stages in the production of a Type I supernova. The process begins with the formation of a binary star system in which one star – the primary – has a mass M_1 greater than that of its companion – the secondary – of mass M_2. Since the primary star is the more massive of the pair it will evolve into a giant star long before the secondary does (Stellar evolution will be discussed in the next chapter). At stage (1) the giant primary star begins to lose mass to the secondary, and the two begin to spiral together within a common envelope. Eventually, the primary giant evolves into a white dwarf star in close orbit about its companion (the now-enhanced mass secondary). Eventually, at stage (2), the secondary star evolves away from the main sequence to become a giant star and mass exchange begins again. At this stage, however, the white dwarf can only accrete so much material before the Chandrasekhar limit is reached. Finally, at stage (3), the white dwarf collapses to produce the Type I supernova and the giant secondary companion is slung outwards at high speed into interstellar space.

Not only this, the atmosphere of such a planet would probably have been badly disrupted long before the material blast wave actually arrived. At distances of less than about 50 to 100 pc, the flux of energetic cosmic rays, combined with that of the X-ray and γ-ray radiation produced by the supernova explosion will probably destroy or at least badly perturb the atmosphere surrounding an Earth-like planet. An intriguing book – *The Cycle of Cosmic Catastrophes* [Bear and Company, Rochester, Vermont (2006)] by Richard Firestone, Allen West, and Simon Warwick-Smith – presents the argument that Earth suffered a triple blow from the supernova responsible for producing the Geminga pulsar and gamma-ray source.[14] Based upon radiocarbon data (and a host

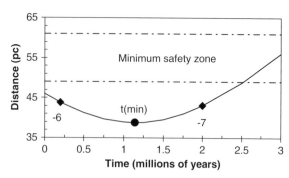

FIGURE 2.12. Variation in the distance to IK Pegasus. The system will be at its closest to the Solar System at a time t(min) = 1.14 million years from now. The safe distance beyond which the supernova will have little effect on Earth's atmosphere is estimated to fall between 50 to 60 pc. The points labeled -6 and -7 correspond to the distances and times at which a supernova will occur when the accretion rate is 10^{-6} and 10^{-7} $M_\odot$/yr respectively.

of other environmental markers), Firestone and co-authors argue that 41,000 years ago the atmosphere encountered the X- and gamma-ray burst from the Geminga supernova event, then 23,000 years later(18,0000 years ago) Earth encountered the supernova shockwave, followed 5,000 years later (13,0000 years ago) by a slower-moving debris wave. Firestone and his co-authors link the various events to mega-fauna extinctions, massive flooding, and dramatic climate change.

The peak energy output per unit time (the luminosity) of a typical Type I supernova is $L_{SN} \sim 10^{36}$ (watts). This is a tremendous amount of energy being radiated into space—indeed, 10 billion times greater than the Sun's present luminosity—and yet it is by no means the most energetic of celestial cataclysms.

The star Eta Carina is the next most likely candidate, among the known stars within a few thousand parsecs of us, to produce a Type II supernova—the result of the catastrophic collapse of a single massive star. Eta Carina is some 5 million times more luminous than the Sun, and it is estimated to be about 120 times larger in size. There is some evidence that Eta Carina is actually a binary system, but it is reasonably clear (whether singular or a binary) that the system contains at least one 50 to 100 $M_\odot$ star. For at least the last century and a half it has undergone

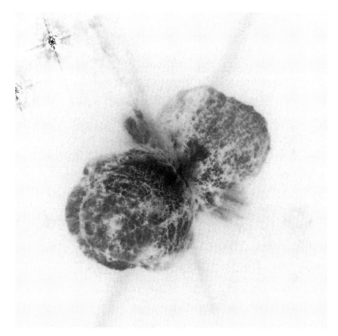

FIGURE 2.13. Eta Carina as photographed by the *Hubble Space Telescope* in the mid-1990s. The twin lobes of the expanding envelope are clearly visible, and the surrounding trelliswork of twisted veins indicates those locations where carbon grains have formed. The envelope is expanding into space at a speed of about 700 km/s. An equatorial disk can also be seen in this image, which has been reproduced in a negative format to enhance the visible detail. (Image courtesy of NASA)

quasi-periodic outbursts in brightness, and since the early 1840s it has lost an estimated $2 - 3\ M_{\odot}$ of material into space. Much of this material resides in a surrounding double-lobed nebula called the Homunculus (Figure 2.13). Various model calculations suggest that Eta Carina will undergo supernova collapse within the next 100,000 years, but since it is situated at a distance some 2,500 pc away from us, the effects of Eta Carina's destruction will be of little consequence to Earth. Indeed, as seen from Earth, the peak brightness of the Eta Carina supernova will be about magnitude -4, comparable in brightness to the planet Venus.[15] In contrast, if IK Pegasus undergoes supernova disruption when it is at its closest point to Earth, it will attain a peak brightness of about magnitude -16, brighter than the full Moon.

GRBs and Hypernovae

By far the most energetic of explosions within the entire universe are those associated with the hypernovae that are responsible for producing gamma ray burster (GRB) events. The GRB production mechanism is believed to result from the gravitational collapse of a rapidly spinning, massive, magnetic star (sometimes called a collapsar; see Figure 2.14). The rotation and the strong magnetic field combine to produce two exceptionally strong and twisted magnetic columns protruding along the star's spin axis. Electromagnetic radiation, such as gamma rays, are then directed along these magnetic columns and literally 'squirted' into space as two oppositely directed jets.[16]

Although the details of the GRB production mechanism are still being worked out, it is clear that they can release a staggering 10^{54} joules worth of energy into space in just a few tenths of seconds. At these energy-release levels a GRB occurring within a few kilo parsecs of Earth could potentially cause biologically significant irradiation. The time interval between GRB events occurring in our galaxy has been estimated to be of order $T_{GRB} \sim 10^8$ years (but see later), and although the exposure time to the burster radiation is just a few seconds (affecting just one hemisphere of Earth,

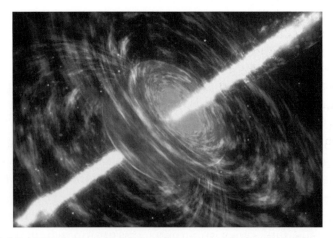

FIGURE 2.14. Artist's image of a collapsar producing a GBR event. The gamma rays are channeled along the oppositely directed, magnetically confined jets. (Image courtesy of NASA)

therefore), the energy flux might nonetheless seriously damage Earth's upper atmospheric structure. One set of calculations by Brian Thomas and co-workers at the University of Kansas[17] finds that even a 10-second exposure to the gamma rays from a GRB 2,000 pc away can severely damage Earth's ozone layer. Indeed, the effect of the gamma rays is to break apart the molecular nitrogen (N_2) in the atmosphere, which then reacts with molecular oxygen (O_2) to produce nitric acid (NO). It is the nitric acid that destroys the ozone (O_3), further producing nitrogen oxide (NO_2). Then, in a feedback loop, the NO_2 reacts with O_2 to produce more NO, which then destroys more O_3. The computer model indicates that it takes at least five years for the atmosphere to recover the ozone lost due to the gamma ray exposure. It is not so much the loss of atmospheric ozone that is the problem for life on Earth, but rather the problem that the Sun's UV radiation, normally absorbed by the ozone, will be able to penetrate to the ground. Extended exposure to solar UV rays will, for example, result in a dramatic increase in animal and human skin cancers; surface dwelling plankton (the bottom of the aquatic food chain) would also be killed off, probably resulting in numerous ecosystem failures.

It has been suggested that the Ordovician mass extinction (see Figure 2.10), in which 60 percent of marine species became extinct some 450 million years ago, was the result of atmospheric disruption caused by a nearby GRB event. The evidence for this, however, is not fully convincing,[18] and recent studies suggest that GRBs must occur well within 1,000 pc of Earth before any serious atmospheric damage is likely to occur. In addition, Krzysztof Stanek and co-workers[19] have recently argued that GRBs tend to occur mostly in small, irregularly shaped galaxies having low heavy element abundances. (These elements will be discussed in the next chapter.) What this means for us is that the odds of any GRBs occurring in our Milky Way galaxy now and in the future are actually very low indeed, and the interval between events is probably even longer than the 10^8 years indicated earlier.

Can anything be done to protect Earth from GRBs? In principle, perhaps surprisingly, yes. A hypernova precursor star will of necessity be highly luminous, and if situated within the range at which damage to Earth's atmosphere might result (i.e., less than 1,000 pc), it will be a very obvious astronomical object. In principle, future inhabitants of Earth will at least know which stars to guard against. Further, detailed spectral observations could also determine the direction of the progenitor star's spin axis, and this would determine if Earth were situated in the direct line of fire. If Earth were actually exposed to a burst of gamma rays from a GRB, then in principle large UV sunscreens could be placed in near-Earth orbit to temporarily protect Earth's surface-dwelling life forms from direct solar UV exposure (Figure 2.15).

The potential importance of hypernovae and GRBs in controlling the emergence of intelligent life has recently been discussed by James Annis.[20] Indeed, Annis argues that our Milky Way galaxy has been sterilized of life on numerous occasions in the past. It is generally believed that the GRB rate decreases with time—that is, GRB events were much more common when our galaxy was younger. Annis argues that because of this effect the

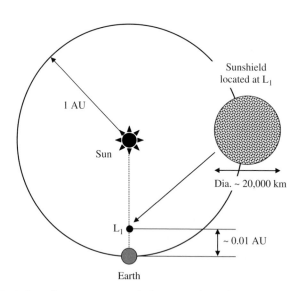

FIGURE 2.15. It has been suggested that Earth's climate might be manipulated by placing a sunshield at the Sun-Earth L_1 point, where gravitational interactions are equal and opposite. Technically, to completely cover the Sun at L_1 the sunshield would need to have a diameter of 20,000 km. Clearly a much smaller diameter sunshield is required in practice, since it is not the intention to block out all of Earth-incident sunlight.

galaxy has only recently undergone a phase-transition allowing for the long-term survival of complex life forms (such as us). His argument runs like this: In the early galaxy GRB events were so common that even if life did manage to evolve on some specific planet it would soon thereafter be destroyed. As the GRB rate decreased with time, however, various strongholds of life could develop and undergo advanced evolution, eventually producing (in some cases) intelligent species capable of space exploration. Annis, in fact, argues that this phase transition requirement offers a solution to Fermi's Paradox. His point is that since intelligent life will not evolve in the galaxy until the GRB rate drops to a level allowing for the phase-transition to occur, the time for the emergence of extraterrestrial civilizations capable of galactic colonization cannot be much different from $T_{US} \sim 4.5$ billion years. Hence, extraterrestrial civilizations are only just on the verge of beginning galactic colonization, and perhaps they will be here very soon.

The Embrace of Andromeda

Of order 20 million Sun diameters separate our Sun from Proxima Centauri. In contrast, only about 20 galaxy diameters separate the Milky Way galaxy from its nearest comparable-sized companion M31—the Andromeda Galaxy. It is because of this basic size compared to separation conditions that star collisions are rare, but galaxy interactions are relatively common. It is certain that the Milky Way galaxy has 'cannibalized' numerous smaller Local Group[21] galaxies in the past with little effect upon its overall structure; the forthcoming interaction with M31, however, will be much more extensive than any of its previous encounters.

The Andromeda Galaxy is presently some 730 Kpc away, but it is moving towards us at a speed[22] of about 120 km/s. We do not know if a direct collision and interaction with our Milky Way galaxy is actually going to occur, but if it does, then at its current speed M31 will merge with our galaxy in something like $T_{M31} \sim 6 \times 10^9$ years (Figure 2.16).

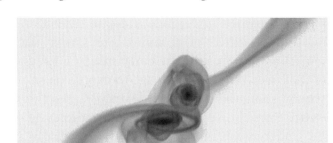

FIGURE 2.16. A snapshot from a computer model simulation of the collision between the Milky Way galaxy (to the upper right) and M31. (Image courtesy of John Dubinski, University of Toronto).

The possible consequences for our Solar System as the merger between M31 and our Milky Way galaxy proceeds are difficult to predict. It is clear, however, that star collisions are still highly unlikely to happen; the average separation between stars will continue to remain at many millions of Sun-diameters. This being said, a close stellar encounter is certainly possible, and our Solar System may yet be disrupted in T_{M31} years from now—intriguingly, a time not far removed from the canonical end of the Sun's main-sequence phase; our distant descendants may be in for double trouble.

Although the star-Sun close-encounter timescale will not be greatly changed by the merger of our galaxy with M31, the time interval T_{NVA} between Type II supernova[12] encounters will decrease. This result follows in the wake of the enhanced star formation that will occur during the M31 collision.[23] The star formation rate will inevitably increase during the M31 merger because of the compression and interaction of the large gas and dust clouds known to reside in the interstellar media of both galaxies.

Our distant descendants will certainly experience a wonderful view as the Andromeda galaxy approaches over the next several billion years. Two Milky Way bands may eventually grace the night sky, and myriad glowing gas clouds, star formation regions and supernovae will sparkle in the heavens. Indeed, a beautiful but potentially deadly vista lies ahead for humanity.

Deep Time

Even if Earth physically survives the inevitable future close encounters with wayward stars, nova explosions, and the ravages of an aging Sun, it still has a 'natural' winding-down mechanism in the form of inescapable gravity. Indeed, any object orbiting around a fixed center will lose energy due to gravitational radiation. It is certainly a small amount of energy that is lost per orbit, but how long it takes is not the issue at this stage. As Earth loses orbital energy through the generation of gravitational waves it will spiral in towards the Sun. The decay time, T_{GRV} will be of order $T_{GRV} \approx (c/V_{orb})^5 \, P$, where V_{orb} is the orbital velocity, c is the speed of light and P is the orbital period. Earth currently has $V_{orb} = 30$ km/s and $P = 1$ year, which dictates a decay time of $T_{GRV} \approx 10^{20}$ years. For the planet Mercury the decay time is $T_{GRV} \approx 10^{16}$ years. Mercury, however, will be consumed by the expanding Sun in about 6 billion years, long before its T_{GRV} destruction time. Earth, on the other hand, may physically survive the Sun's red giant phase, and correspondingly, some 8 billion years from now find itself in orbit around a white dwarf star. If it continues to survive against disruption through close stellar encounters, then its eventual demise (in the deep, deep future) will be to spiral into the surface of a zero temperature, zero luminosity black dwarf sphere[24] supported by degenerate electron pressure—the cold and dark relic of the Sun.

The Doomsday Event

The great philosopher and biologist J. B. S. Haldane once remarked that "The universe is not only as queer as we suppose, but queerer than we can suppose." Haldane is indeed right, and it is entirely possible that there are physical entities within our Milky Way galaxy that might destroy Earth (and the Solar System) in its entirety if encountered just the once. Such entities might include massive black holes, putative strange matter, and of course, the phenomena that Haldane tells us we know nothing about. More will be said about black holes in Chapter 5, but since there is currently no consensus on what the mass distribution of black holes is, we can presently say very little about the chances of Earth or the Solar System encountering one (other than it hasn't happened in 4.5 billion years). Strange matter is certainly odd stuff, but entirely possible within the context of allowed modern atomic physics. The suggestion is that such matter, made of strange quarks (the fractionally-charged, basic building blocks of matter), might form within the interiors of high-density neutron stars, and it may also be stable (in the form of so-called stranglets) outside of such objects. It has been further speculated that should ordinary matter (such as the stuff Earth is made of) meet a stranglet, then it will become transformed into strange matter. There is currently no proof that strange matter exists, but Haldane's comment surely resonates on such issues.

The fact that Earth formed some 8 or 9 billon years after the universe came into existence (the moment of the Big Bang) and has survived for at least 4.5 billion years tells us that doomsday catastrophes must be rare in our galaxy, but the question is, how rare? If a randomly occurring event (i.e., a stranglet or massive black hole encounter) destroys a planetary system at some constant rate tau, τ, then the probability that a specific planetary system will survive a time t decreases as $e^{-t/\tau}$. That is, the survival probability decreases exponentially with time. If the event causing destruction occurs relatively often, then τ will be small and the survival probability soon becomes very small. If destruction events are rare, then τ will be large and the probability of a planetary system surviving for a long time is high. It is not, unfortunately, possible to constrain τ with any certainty at the present time, other than by making the statement that the continued existence of the Solar System suggests that it is unlikely that tau is less than several billion years.

The Long and the Short of It

The timescales upon which astronomical disasters are likely to occur are compared in the inequality below.

$$T_{NEA}(1-\text{km}) << T_{NVA} \approx T_{LPC} \approx T_{GRB} < T_{M31} < T_{MS}$$
$$<< T_{CE}(d=5\text{AU}) << T_{GRV}$$

At the far right hand end of the time sequence, if nothing else destroys Earth first, it will meet its ultimate doom by being accreted onto the surface of the Sun, which will have evolved into a black dwarf star. In the near term, however, the most likely astronomical disaster would be that due to an impact by a relatively small (perhaps 100 to 1,000 meters across) asteroid or comet. These are the smaller, more common near-Earth objects (NEOs) that feed into the inner Solar System from the main belt asteroid region and the Jupiter family of comets. The most serious impacts, however, will be those from the long-period comets. On about the same timescale that a cometary shower impacts the Solar System, it is liable to be irradiated by a near-by supernova outburst and possibly by a GRB event.

Although Annis[20] has argued that GRB events are capable of sterilizing the galaxy, not all researchers agree with this claim. Indeed, it has been pointed out that just a few centimeters of rock, soil, or water would completely protect any life forms on the Earth's surface from GRB irradiation.[25] Perhaps the most important issue with respect to surviving a GRB event is the time required to regenerate atmospheric ozone (destroyed by the incoming γ-rays). It is the ozone shield that protects life on Earth from the harmful UV solar-background radiation.[21]

Making the Best of It

The first steps toward securing humanity's long-term survival will probably focus on the means of detecting and then either destroying or diverting potential Earth-impacting asteroids and comets. Methods will also have to be developed for repairing Earth's atmosphere following the ravages imposed by supernovae and GRB irradiation events. Although each of these catastrophe-avoiding steps will require great ingenuity, enormous engineering skill, tremendous daring, and total cooperation between the world's nations, there is in principle no specific or fundamental reason why they cannot be achieved. Time, of course, will tell how well humanity lives up to the task. What is perhaps most important to note at this moment, however, is that there is a straightforward progression in project size and required skill development that takes us from impact avoidance to terraforming, Solar System colonization, and ultimately to solar rejuvenation. Once again, the deep future is not isolated from our present.

Before moving on to consider the detailed properties of the Sun and how it might be rejuvenated, let's briefly review a few of the larger-scale, stepping-stone projects that our descendants may eventually want to undertake.

Dyson Spheres

The Sun radiates a total of 3.85×10^{26} joules of electromagnetic energy into space per second. At a distance of 1 astronomical unit from the Sun, Earth receives a flux of 1,369 watts/m^2 of solar energy at the top of its atmosphere. Multiplied by Earth's cross sectional area of 1.3×10^8 km^2, this translates to an energy budget of about 1.8×10^{11} watts. Earth, therefore, taps only a minuscule amount of the Sun's energy output, something like 3.4×10^{-17} percent of the available energy, in fact. Indeed, small fry. In what has become a classic paper, Freeman Dyson in 1960 proposed that the existence of extraterrestrial civilizations might be revealed through the infrared emission from artificially constructed worlds.[26] Specifically, Dyson argued that a sufficiently advanced civilization might utilize the material extracted from a Jovian-type planet to engineer a sphere around its parent star. The engineered shell in our Solar System might have a radius of about 1 AU and be several meters thick. In principle this shell could support extended human colonies and large industrial complexes capable of tapping a substantial fraction of the Sun's energy output. The heat signature of the shell (now commonly called a Dyson sphere) would be that of a blackbody radiator with a temperature of 200 to 300 K. Such structures would likely be bright infrared sources in the 10-micron wavelength region of the electromagnetic spectrum. Interestingly, and unlike the radio surveys that search for intentional beacons or leaked transmissions from extraterrestrial civilizations, Dyson-like spheres would be potentially observable without the specific intent of the originating civilization.

Dyson spheres have indeed been searched for, but no finds have been reported to date.[27] The problem, however, with serendipitous detection surveys (just as with the radio search for extraterrestrial signals) is to know when a null result actually indicates that there are no sources. At present we cannot say, one way or the other, whether any galactic civilization has built Dyson sphere-like structures. Clearly, engineering ability aside, there is a cost versus potential payback tradeoff that any civilization has to consider before embarking on the construction of a Dyson sphere. It may, in fact, make more sense to engineer the smaller real estate within a planetary system first. For example, within our Solar System, the terraforming of Mars and Venus might represent the first steps in this direction.

Terraforming

Vast amounts of literature have been published on terraforming—literally, the engineering of an otherwise barren world into one compatible with biological life. As a basic minimum requirement, however, the process of terraforming should result in the presence of liquid water on the surface of the planet being engineered. Liquid water will only exist on the surface of a planet if the average surface temperature (Figure 2.17) is above 273 K and if the atmosphere provides a surface pressure in excess of 610 Pa (which defines the so-called triple-point condition). In the case of Mars and Venus, the would-be engineer faces a range of problems—problems that, in fact, fall at the two extremes of the terraforming spectrum. Mars is a frozen, geologically inactive world with a tenuous atmosphere, while Venus is (possibly) geologically active, has a massive atmosphere, and a surface temperature in excess of 700 K. For Mars, the atmospheric pressures and temperatures are too low to allow liquid water to remain very long on its surface, while on Venus the surface temperature is far too hot and the atmospheric pressure is 90 times greater than that on Earth.

Mars is a relatively small planet, about half the size of Earth, and its interior consequently cooled off rapidly, resulting in the permanent shutdown of the majority of its volcanic networks.[28] This latter effect is particularly important with respect to the

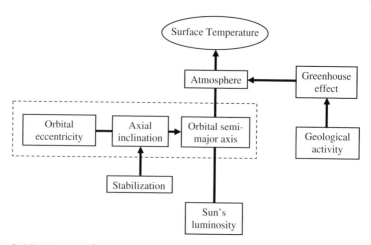

FIGURE 2.17. Factors determining the surface temperature of a planet. The irradiance by the Sun will depend upon the planet's orbital characteristics, the planet's geological activity, and the structure and composition of its atmosphere. The variation in the orbital and axial tilt parameters (dashed line box) are governed by the Croll-Milankovitch cycle in the case of Earth.[6]

heating of the planet's surface, since it is the geochemical carbon cycle that normally controls the greenhouse gas content of a planet's atmosphere. On Earth, for example, chemical weathering and plate tectonic activity control the cycling of carbon between gaseous and mineral phases. The volcanoes formed at the plate subduction zones allow for the release of carbon dioxide (CO_2) into the atmosphere, thus increasing the greenhouse heating effect. At the same time the CO_2 in the atmosphere is chemically incorporated into rocks such as limestone by the weathering and erosion of silicate rocks, once again trapping the carbon in a solid phase. It is the feedback loop between the temperature-sensitive chemical weathering process (it runs more rapidly at warmer temperatures) and the release of CO_2 by volcanic activity that has allowed Earth to maintain a near constant temperature for the past several billions of years (Figure 2.18). On Mars, the CO_2 release mechanism via volcanic outgassing no longer works and, consequently, its ancient atmosphere has literally frozen out. The CO_2 is now mostly trapped within its constituent rocks.

For the surface temperature of Mars to once again rise above 273 K the CO_2 content of its present atmosphere would have to be increased by a factor of about 200. This would raise the greenhouse

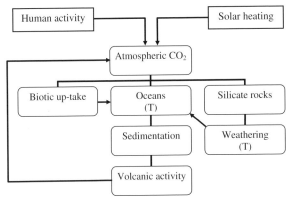

FIGURE 2.18. The terrestrial carbon-dioxide cycle. The (T) term in the Oceans and Weathering boxes indicate specific temperature sensitivity. If, for example, the oceans are warm, then carbonate minerals are more rapidly formed, and the rate at which the ocean takes up CO_2 increases concomitantly, reducing the atmospheric CO_2 concentration. The reduced atmospheric CO_2 content then results in global cooling, which eventually checks the ocean-warming effect. Likewise, an increase in atmospheric temperature increases the rate at which silicate rocks can dissolve CO_2 to produce carbonate rocks. The hotter the temperature, the faster the 'weathering' effect on silicate rocks, and the greater the rate at which carbonates are deposited into the ocean. The CO_2 contained in the sedimentary carbonate rocks is eventually returned to the atmosphere through volcanic outgassing. Human activity is presently causing an increase in atmospheric CO_2 through, for example, the burning of fossil fuels and the clear-cutting of ancient forests. The atmosphere is directly heated by the Sun, and this heating effect will increase with time, eventually forcing – if left unchecked – the loss of Earth's oceans and the death of the biosphere.

gas content to a level that it could offset the low solar irradiance at the orbit of Mars. Such an increase in the CO_2 abundance would also raise the surface pressure well above 610 Pa, allowing liquid water to exist in a stable form at its surface. In principle such a process could be initiated by manipulating (that is, increasing) the brightness of Mars with the aid of space reflectors.[29]

Although the atmosphere of Mars needs to be warmed up by the terraforming engineer, the atmosphere of Venus needs to be cooled down. This process could in principle be achieved by reducing the amount of CO_2 in the Venusian atmosphere, either by physically removing it or by increasing the CO_2 uptake by the planet's crust. Radiation shields or reflectors would also need to be installed in orbit around the planet to reduce the solar irradiance.[30]

Indeed, even if there were no greenhouse heating factor for Venus, its surface temperature would be about 313 K (that is, an uncomfortably high 40°C) based purely upon its orbital proximity to the Sun.

Transforming Mars and Venus into habitats suitable for nurturing biological life will only be achieved through the investment of considerable money and through long-term cooperation between all nations on Earth. It would also require the development of the engineering skills needed to perform such large-scale operations. Some of these skills will, with little doubt, first be honed in the engineering of Earth's atmosphere. Global warming resulting from the generation and release of greenhouse gases, wide-scale industrial pollution, along with poor land use and agricultural management, are all conspiring to destroy the natural feedback systems that have maintained Earth's relatively steady global temperature over the eons. Professor James Lovelock has introduced the profoundly important Gaia concept to describe the whole-Earth feedback system and, in his recent book *The Revenge of Gaia: Why Earth Is Fighting Back—and How We Can Still Save Humanity* (Penguin Books, London, 2006), he paints a disturbing picture for our near-term future. Human activity is changing Earth's atmosphere and there are clear signs that traditional weather patterns have changed and will continue to change for the worse. Stronger and more numerous hurricanes (Figure 2.1) and tropical storms, the melting of Arctic and Antarctic ice fields, the retreat of glaciers around the world—all are being observed.[31] Humanity's short-term (over the next few hundred years) survival may well depend upon the engineering skills that our children develop in order to revitalize Gaia and Earth's atmosphere.

One suggestion that has been made for solving the global warming problem calls for the injection of sulphur dioxide (SO_2) into Earth's atmosphere to increase its albedo, thereby causing more solar energy to be reflected back into space.[32] This, in fact, mimics the cooling effects associated with the sulfur plumes emitted by volcanoes.[33] Another possible solution is to increase the absorption of atmospheric CO_2 in the oceans by seeding surface waters with iron particles.[34] Physicist Edward Teller and co-workers[35] have further suggested that the global warming effects due to enhanced atmospheric CO_2 concentrations could be counteracted by the deployment of an electrically conducting mesh in Earth's stratosphere (or in near-Earth orbit) with the aim of reducing the direct solar heating of the atmosphere. Alternatively, Jerome Pearson and co-workers[36] have proposed that atmospheric insolation could be increased by engineering an equatorial ring of particles around Earth (similar to Saturn's rings). Indeed, by making the ring particles out of dismantled near-Earth asteroids (NEAs), this proposal also solves some of the potential impact problems faced by our descendants.[37]

In an ideal world the drastic manipulation of Earth's atmosphere would not be required. Gaia currently does it for free. However, given the apparent lack of political leadership on industrial emissions issues, such as those recommended in the Kyoto Accord, it is not at all likely that the basic engineering skills needed to terraform Mars and Venus will be developed on Earth during the next 100 years. Although we already have the ability (by default) to alter planetary atmospheres (that is, terraform) by the careless exploitation of Earth's resources, what is needed in the future is the ability and skill to control the alterations.[38] Lovelock poignantly writes in his latest book, however, "The idea that humans are yet intelligent enough to serve as stewards of Earth is among the most hubristic ever." Humanity has much to learn and there is much work that needs to be done.

Space Structures

That our descendants will eventually exploit all of the useful and available real estate within our Solar System seems inevitable. Terraformed Mars and Venus, enclosed Moon and Mercury bases, industrial processing plants located in the main asteroid belt and on the Jovian Moons—all may come about. Part and parcel with this development will be humans adapting to living in space. *Skylab*, *Mir*, and the *International Space Station* have already supported small (non-reproducing) human colonies, but in the future kilometric-scale space structures, such as those envisioned by physicist Gerard O'Neill, will presumably be engineered.[39] Indeed, the space structures considered by O'Neill are truly grand in conception. Composed of 30-km long x 6-km diameter rotating cylinders, the space platforms would have an Earth-like atmosphere (and weather system) and potentially house, feed, and employ 10,000 people. The challenge, as noted by O'Neill,[39] "is to bring the goal of space colonization into economic feasibility now, and the key is to treat the region beyond Earth not as a void but as a culture medium, rich in matter and energy."

In addition to building space colonies and terraforming, Paul Birch has suggested that rigid shells might eventually be engineered around all of the planets (and some of their moons) within the Solar System.[40] Such supramundane planets would have habitable atmospheres, natural gravity, and be made of material extracted from asteroids, smaller planetary moons, and even the Sun. Perhaps somewhat overly optimistic, Birch suggests that the first supramundane habitats will be built around the planet Venus by 2040!

Thinking Long-Term

The solution to Fermi's Paradox that is being followed in this book, as outlined in Chapter 1, is that advanced civilizations do not engage in galactic colonization because they have found no need to. This is not to suggest that stellar exploration won't ever take place. But a clear distinction between local exploration and galactic colonization is drawn. Perhaps the first stellar voyages will be initiated in order to divert nearby rogue stars that are heading for an uncomfortably close encounter with the Solar System [i.e., on a timescale of T_{CE} (d = 5000 AU) $\approx$ 200 million years, or T_{M31} ~6 billion years]. Indeed, the possibility of controlling the space motion of a star through the induction of mass loss has been discussed by David Criswell[41] (an operation described in more detail in Chapter 5). Long-time advocate and researcher into terraforming and space colonization Martyn Fogg[42] has further investigated the possibility of star exchange to ensure the long-term habitability of the Solar System.

It does not seem unreasonable to assume (or at least adopt as a working hypothesis) that all intelligent life forms that are capable of molding and adapting their environments will expand to live upon and commercially exploit all of the accessible components of their parent planetary system.[43] In addition, it also does not seem unreasonable to assume that all advanced civilization will find some way of avoiding a Malthusian catastrophe[44] in which the demands of a population overreach the available food and raw materials supply. Indeed, our descendants will have to solve this problem if they are to have a long-term future. In addition, the long-term survival of a civilization (in, say, a planetary system similar to our own) will require the development of planetary defense programs to shield against asteroid and comet impacts. Such programs will, in fact, probably constitute the first truly global space initiatives, literally (certainly hopefully) uniting humanity through a common goal for survival against otherwise devastating odds. By tackling the NEO impact problem, as a space-based initiative, humanity will begin to develop the skills necessary to tackle even larger-scale projects.

Again, as a working hypothesis, it does not seem unreasonable to assume that all advanced extraterrestrial civilizations will undertake terraforming projects, construct colonies and outposts on planetary moons, and engineer O'Neill-style spacecraft to produce their food. All such projects will anchor a civilization to its planetary system. Its financial investments, its sense of self, and indeed all that a civilization might physically require to exist will be infused within its home planetary system. Such an investment of time, lives, and money is hardly something that any civilization will willingly abandon and move away from. Indeed, such a holistic system is worth saving by any and all means that fall within the realms of possible physics and engineering.

In terms of timescale the first astronomical danger that our descendants will need to guard against is that from NEO impacts ($T_{NEO} \sim 10^{5-6}$ years), then long-period comet impacts and supernova irradiation ($T_{LPC} \sim T_{NVA} \sim 10^{6-7}$ years). The next significant timescale – indeed the ultimate timescale of concern in this book – will be the exhaustion of hydrogen in the Sun's core about 5 billion years from now. If Earth and our descendants are to survive beyond the canonical T_{MS} then two options present themselves: either tame the natural gigantism of the aging Sun by rejuvenating it, or leave the Solar System for good. Our descendants may be the first to reach for the stars in a quest for a new Earth, leaving a devastated and blistered Solar System (as described in Chapter 4) far behind them. But, as shall be explained later in Chapter 5, it is not unreasonable to think of extending the lifetime of the Sun, and thereby the potential life-supporting viability of the Solar System, by a factor of between 10 and 15, to T_{MS}(engineered) $\sim 10^{11}$ years. This engineered lifetime is still three orders of magnitude smaller than the planet scattering encounter timescale by a nearby rogue star [i.e., T_{CE} (d = 5 AU) $\sim 2 \times 10^{14}$ years], but it is surely better than doing nothing. Either the Sun becomes a giant some 5 billion years from now and in the process destroys all life within the Solar System, or through the act of rejuvenation our descendants create an additional 95-billion-year time span over which to enjoy the nurturing light and warmth of a tamed Sun.

Notes and References

1. Strictly speaking there are small amounts of deuterium and lithium produced during the time of primordial nucleosynthesis. Interestingly, the fusion of deuterium to produce ^{3}He (via the reaction: D + P $\Rightarrow$ ^{3}He + gamma ray) is of great importance during the early (pre-main sequence) stage of star formation. Some of the consequences of stars undergoing an early deuterium burning phase are discussed in M. Beech and R. Mitalas, The formation of massive stars, *Astrophysical Journal Supplement*, **95**, 517–534 (1994).

2. A. S. Eddington, *The Internal Constitution of the Stars*, Cambridge University Press, Cambridge (1926). Although the physical arguments presented in this book are now somewhat dated, they remain a brilliant example of lucid science writing.

3. K. Yau, P. Weissmann, and D. Yeomans, Meteorite falls in China and some related human casualty events, *Meteoritics*, **29**, 864–871 (1994) provide a summary of meteorite falls recorded in Chinese chronicles between 700 B.C. to A.D. 1920. They report on one event that occurred in Ch'ing-yang in 1490, where it is claimed that stones fell like rain and many tens of thousands of people were killed. The reported death toll seems very high, but this may possibly describe a Tunguska-like impact (see Note 5 below) event where meteorites actually survived to hit the ground (and people). As shown in the author's book *Meteors and Meteorites: Origins and Observations* (Crowood Press, 2006, p. 26), the probability of a person who lives to the ripe old age of 99 years being struck by a 1-gm meteorite is one

chance in 1.5 billion. The other way of saying this is that one person should be hit (not necessarily killed) worldwide every 33 years (or so).

4. Three NEA groups are generally recognized: the so-called Apollo, Amor, and Aten asteroid groups. Both the Apollo and Amor groups have their aphelia (greatest distance from the Sun) within the main belt asteroid region. The Apollo NEA group has perihelia (closest point to the Sun) inside of Earth's orbit, while the Amors have perihelia just outside of Earth's orbit. The Aten NEA group have perihelia well inside and aphelia just outside of Earth's orbit.

5. It has been estimated that the equivalent of some 5 megatons of TNT energy was liberated in the Tunguska explosion. In the comprehensively titled research paper, Earth impact effects program: a web-based computer program for calculating the regional environmental consequences of a meteoroid impact, by Gareth Collins, Jay Melosh and Robert Marcus [*Meteoritics and Planetary Sciences*, **40** (6), 817–840 (2005)], the equations that describe the environmental effects of an impact (such as crater size, overpressure, thermal radiation, blast wave propagation, and seismic effects, as well as casualty levels) have been gathered together. The web site is accessible at http://www.lpl.arizon.edu/impacteffects.

6. Not every mass-extinction event observable in the fossil record can be associated with an impact. Some are, without a doubt, geological in nature and relate, for example, to variations in sea-level, anoxic ocean events, extensive volcanism, tectonic activity as well as climate change [see the excellent book by Tony Hallam, *Catastrophes and Lesser Calamities—The Causes of Mass Extinctions*, Oxford University Press, Oxford, 2005]. Impact extinction events can be identified by a significant increase in the iridium abundance in the terrestrial rocks that indicate the extinction boundary. Comets and asteroids are relatively rich in iridium (compared to Earth's crustal rocks) and, following a large impact event, the constituent iridium is deposited over much of Earth's surface, forming a thin iridium-rich layer. Perhaps the most famous such iridium rich boundary layer is that which was produced by the impact that resulted in the 180-km diameter Chicxulub crater in the Yucatan peninsula at the end of the Cretaceous Period (65 million years ago). On a smaller scale Jan van Dam of Utrecht University in the Netherlands, and co-workers [Long-period astronomical forcing of mammal turnover. *Nature*, **443**, 687–691 (2006)] have found that there are 'spikes' in rodent extinction record spaced at intervals of 2.4 million and 1.2 million years. These intervals, van Dam argues, correspond to variations in Earth's orbit and shifts in the tilt of Earth's rotation axis to its orbital plane. The

cycles of extinction have apparently been active for at least the past 22 million years. The Croll-Milankovitch cycle, which links long-term climate change to variations in Earth's orbit, is generally believed to control the timing of Earth's ice ages; events that invariably lead to many species becoming extinct [for a description of the C-M cycle see, Doug Macdougal, *Frozen Earth: The Once and Future Story of Ice Ages.* University of California Press, Berkeley (2006), pp. 65–88].

7. Simply blasting an asteroid with a volley of, say, nuclear missiles may actually increase the impact devastation. Instead of one object hitting Earth, a series of explosions might result in multiple fragments striking Earth over a much larger area than that corresponding to a single-body impact. This is especially a problem given the current understanding that many asteroids are 'rubble piles' of loosely bound components. A good general review of the observational status and future spacecraft missions to comets and asteroids can be found in the article by A. C. Levasseur-Regourd, E. Hadamcik, and J. Lasue, Interior structure and surface properties of NEOs: what is known and what should be understood to mitigate potential impacts. *Advances in Space Science*, **37**, 161–168 (2006).

8. A space tractor concept is discussed in some detail by E. Lu and S. Love in their article, Gravitational tractor for towing asteroids, *Nature*, **438**, 177–178 (2005). The European Space Agency (ESA) is currently considering a spacecraft mission (the Don Quijote mission) to study the internal structure of an asteroid (presently unspecified). The mission will also attempt to change the orbit and rotation state of the target asteroid through high-speed surface impacts. An overview of the ESA mission can be found in a 2006 paper presented by Ian Carnelli, Andres Galvez and Dario Izzo, Don Quijote: a NEO deflection precursor mission, to a NASA workshop dedicated to near-Earth object detection, characterization, and threat mitigation— available at: http://www.esa.int/gsp/ACT/doc/ACT-RPR-4200-IC-NASANEOWS-DonQuijote.pdf.

9. The Oort Cloud, named after the Dutch astronomer Jan Oort (1900–1992) who first suggested its existence, delineates the outer edge of the Solar System. It is literally the three-dimensional boundary beyond which the Sun's gravitational influence no longer holds sway over the rest of the galaxy. It is a dynamical boundary that expands and contracts according to the time dependent distribution of matter in the solar neighborhood. Of order 10^{12} to 10^{13} cometary nuclei are believed to delineate the Oort Cloud region stretching between ~10,000 to ~50,000 AU from the Sun. Cometary nuclei within the Oort Cloud region are continually ejected into

interstellar space and into the inner Solar System as a result of gravitational perturbations from close passing stars, massive molecular clouds embedded within the interstellar medium, and during spiral arm region crossings.

10. J. Garcia-Sanchez et al., Stellar encounters with the Solar System, *Astronomy and Astrophysics*, **379**, 634–659, (2001). In addition, R. A. Matthews [The close approach of stars in the solar neighborhood. *Q. J. R. Astr. Soc.* **35**, 1–9. (1994)] finds that Proxima Centauri will not be at its closest approach to the Sun for another 26,700 years. At that time it will be 0.941 pc away.

11. See, for example, the recent research paper by W. M. Napier, Evidence for cometary bombardment episodes, *Monthly Notices of the Royal Astronomical Society*, **366**, 977–982 (2006). A more general account of cometary impacts is given in the very readable book by Duncan Steel, *Target Earth: The Search for Rogue Asteroids and Doomsday Comets That Threaten Our Planet* [Readers Digest Association Inc., New York (2000)].

12. D. H. Clark, W. H., McCrea, and F. R. Stephenson [Frequency of nearby supernovae and climatic and biological catastrophes, *Nature*, **265**, 318–319 (1977)] investigated the possibility of Earth encountering the radiation burst or blast wave associated with a Type II supernova (see Note 13 below) during galactic spiral arm crossings. They concluded that such an 'encounter' might take place once every 10^8 years, assuming a Type II SN rate of one every 100 years. The current, revised SN rate is actually some three to four times higher than that adopted by Clark et al., with one SN occurring every 25 to 30 years. This brings the typical time between close SN passages to a value of order 3×10^7 years.

13. Two main categories of supernova are generally recognized. Type I novae are produced through the accretion-driven collapse of a white dwarf star in a binary system. A Type II supernova is produced, on the other hand, by the collapse of the iron-rich core of a massive $(M_{initial} > 8 \ M_\odot)$ star. The supernova types are generally distinguished observationally according to their spectra, brightness variations with time, and galactic location – Type II supernova, for example, only occur in spiral arm regions where massive stars are actively forming.

14. A pulsar is a rapidly spinning neutron star, and it is believed that such objects form during the supernova disruption of massive stars (i.e., in Type II SN). Spinning at a rate of 4.2 revolutions per second, the Geminga pulsar is believed to be about 300,000 years old. While Firestone and co-authors have assembled an impressive quantity of data to support their thesis, it should be pointed out that much of

what they claim is controversial and almost all their data has multiple possible interpretations not necessarily requiring external (that is astronomical) driving influences.

15. Optical astronomers measure stellar brightness in terms of apparent magnitude. The apparent magnitude of a star is based upon the logarithm of the energy flux received from the star at the surface of Earth. The magnitude scale works in such a way that the brighter an object is, the more negative its magnitude. The faintest star visible to the human eye has an apparent magnitude of about +6, while the Hubble Space Telescope can record stars and galaxies as faint as magnitude +24. In contrast, the full Moon has an apparent magnitude of –12, and the apparent magnitude of the Sun is –27.

16. Collapsars are believed to be the final stage of massive star evolution. The progenitors are the highly distinctive Wolf-Rayet stars, first studied by the French astronomers Charles Wolf and Georges Rayet in the early 1900s. These stars are distinguished by showing emission lines in their spectra (virtually all other stars show only absorption lines) and they undergo extreme mass loss. By way of comparison, the Sun loses mass via a stellar wind at a rate of $\sim 10^{-14}$ $M_{\odot}$/yr, while a Wolf-Rayet star loses mass at a rate of 10^{-4} to 10^{-5} $M_{\odot}$/yr. The Wolf-Rayet stars are believed to represent the end stages of the evolution of stars with initial masses greater than ~ 30 $M_{\odot}$, and the very final hypernova collapse is believed to produce a black hole rather than a neutron star. Working with Professor Romas Mitalas at the University of Western Ontario, this author studied the effects of mass loss upon massive stars [Effect of mass loss and overshooting on the width of the main sequence of massive stars, *The Astrophysical Journal*, **352**, 291–299 (1990)] and found that the mass loss strongly reduces the observed luminosity but increases the surface temperature—a result that actually proves useful with respect to star-engineering. An alternative model for the cause of hypernova and GRB generation envisions the coalescence of two neutron stars previously orbiting each other in a binary system. Such mergers will again probably result in the formation of a black hole. Michael Shara discusses the possible consequences of star collisions and mergers in When stars collide, *Scientific American*, **287** (5), 46–51 (2002).

17. Brian Thomas et al., Terrestrial ozone depletion due to a Milky Way Gamma-Ray burst. *Astrophysical Journal*, **622**, L153–L156 (2006). The authors conclude that a 10-second burst of gamma rays results in a globally averaged ozone depletion of order 35 percent (with some latitudes seeing a 55 percent reduction). A 50 percent decrease in

the ozone column translates to a three-fold increase in the UVB flux received at Earth's surface.

18. J. Scalo and J. C. Wheeler [Astrophysical and astrobiological implications of gamma-ray burst properties. *Astrophysical Journal*, **566**, 723–737 (2002)] have discussed the potential biological consequences of Earth encountering a strong burst of gamma rays. They point out that very little gamma-ray radiation actually reaches Earth's surface and, indeed, a column of water just a few tens of centimeters thick would fully protect any marine organisms from a dangerous dose of radiation. Shielding by small amounts of rock and soil would protect other organisms.

19. Krzysztof Stanek et al, Protecting the Milky Way: metals keep the GRBs away. Paper submitted to the *Astrophysical Journal* in April 2006.

20. J. Annis. An astrophysical explanation for the Great Silence. *Journal of the British Interplanetary Society*. **52**, 19–22 (1999).

21. Our Milky Way Galaxy is part of the so-called Local Group (LG) of galaxies. The LG contains at least 36 members within a region that stretches about 1 Mpc across its longest dimension. Most of the LG members are dwarf elliptical and irregular galaxies with masses of between one thousandth to one millionth that of the Milky Way. The Andromeda Galaxy (M31) is comparable in size and mass to our galaxy. Detailed studies of the central region of M31 reveal a double nucleus indicative of a past 'collision' and 'disruption' of a smaller LG member by Andromeda.

22. In order to determine the true spatial velocity and direction of motion of M31 a measure of both its line-of-sight (radial) velocity as well as its velocity tangential to the line-of-sight is required. Only the radial component is known and, consequently, it is not clear under what circumstances M31 will actually 'collide' with the Milky Way.

23. The burst of star formation triggered by the (potential) merger of M31 with the Milky Way will certainly produce large numbers of massive stars (stars with initial masses greater than 8 $M_\odot$). These massive stars have lifetimes of just a few million years (as indicated by Equation 2.1), and they eventually undergo core-collapse as Type II supernovae. Type II supernovae are typically less energetic than Type I supernovae (see Note 13 above), but the starburst conditions should result in a far greater number of such events occurring in our galaxy than observed at the present.

24. A black dwarf is the name given to the cold (near zero degrees Kelvin) remnant of a fully cooled-off white dwarf. Such objects remain stable against collapse due to the electron degeneracy of their interior, the

pressure support for such gases being independent of temperature. The cooling time required to produce a black dwarf, however, is longer than the present age of the universe.

25. J. Scalo and J. C. Wheeler (see Note 18) point out that any life forms that might have evolved on Mars, even with its higher density atmosphere of the past, would not have fared so well as terrestrial organisms. They estimate that since the formation of the Solar System some 1,000 biologically significant GRB-related events have possibly occurred, and that surface life on Mars (i.e., eukaryotic bacteria) has probably been 'sterilized' many times over. This being said, Professor Alexander Pavlov (University of Arizona) and co-workers have recently argued [Was Earth ever infected by Martian Biota? Clues from radioresistant bacteria. *Astrobiology*, **6** (6), 911–918 (2006)] that the radiation tolerance of some terrestrial bacteria (e.i., *Deinococcus radiodurans*) is more likely to have evolved on Mars than Earth. The bacteria, having evolved on early Mars, were later transported to Earth via Martian meteorites.

26. Freeman Dyson, Search for artificial stellar sources of infrared radiation, *Science*, **131**, 1667 (1960). A shell of radius 1 AU has a surface area of order $2 \times (4 \pi) \times (1.496 \times 10^8)^2 = 5.6 \times 10^{17}$ square kilometers. The factor of 2 accounts for both the inner and outer surfaces of the shell. Compared to Earth's present land area, this corresponds to a 4 billion-fold increase in potential lebensraum. This being said, the shell of a Dyson sphere cannot, for dynamical stability reasons, be solid, but must consist of a 'swarm' of moonlets. Richard Carrigan, Jr. [Searching for Dyson spheres with Planck spectrum fits to IRAS, *International Astronomical Congress paper: IAC-04-IAA-1.1.1.06*] provides an extensive review of observational searches. In a Phase I study of some 4,400 infrared point sources detected with IRAS (the Infra-Red Astronomical Satellite flown in 1983), Carrigan finds that about 1 in 600 have flux characteristics consistent with the radiation expected from a pure blackbody radiator with a temperature in the range $150 < T(K) < 500$ [see, http://home.fnal.gov/~carrigan/Infrared_Astronomy/Fermilab_search.htm].

27. It is not absolutely clear that Mars is volcanically dead. The crystallization ages of a number of Martian basaltic meteorites are as young as 165 million years. This corresponds to about 4 percent of the planet's age. As pointed out by William Hartmann [*A Traveler's Guide to Mars*, Workman Pub. NY (2003)], it is unlikely that Mars has been volcanically active for the past 96 percent of its history and then suddenly, in recent times, all eruptions stopped.

28. Solar reflectors can be used to either enhance the heating of a planet by reflecting additional sunlight into its atmosphere, or they can act as cooling shades by directly blocking the incoming sunlight. It has been proposed that the effects of CO_2 global warming on Earth could be offset by placing a 2,000-km diameter parasol between Earth and the Sun [see, for example, Martin Hoffert and co-workers, Advanced technology paths to global climate stability: energy for a greenhouse planet, *Science*, **298**, 981–987 (2002)].

29. The U.S. National Academy of Sciences published a policy report in 1992 in which it was suggested that the effects of CO_2 warming could be counteracted by installing 55,000 mirrors – each of surface area 100 km^2 – in random inclination, near-Earth orbits. The effect of the mirrors would be to reduce the amount of solar energy intercepting Earth's atmosphere. A similar concept could be used to reduce the direct solar heating of Venus.

30. The U.S. Climate Change Science Program (www.climatescience. gov/) released a report in April 2006 concluding that over the past 50 years there is clear and uncontroversial evidence for human activity-induced climate change. Likewise, Gabrielle Walker [The tipping point of the iceberg, *Nature*, **441**, 802–805 (2006)] describes in sobering detail the distinct climate changes being recorded in Earth's polar regions. It is perhaps a little ironic that some of the effects of human-induced global warming have been masked by anthropogenic aerosol production, which tends to actually cool the planet [see, for example, Jim Coakley's commentary article, Reflections on aerosol cooling. *Nature*, **438**, 1091–1092, 2005]. The irony of this situation is that as some (indeed a sad, small few) nations begin to curb and set new limits on the industrial emission of aerosols, the cooling effect that the aerosols cause will be lost—a process that will actually enhance global warming.

31. F. J. Dyson and G. Marland, Technical fixes for the climatic effects of CO_2. Workshop on the global effects of carbon dioxide from fossil fuels. *DOE report CONF-770385* (1979).

32. The 1991 eruption of Mt. Pinatubo induced a global mean temperature cooling of $\sim$0.5 K as a result of the release of sunlight-scattering SO_2-based particles. The dispersal of SO_2 into the atmosphere would, in principle, increase Earth's insolation by increasing the amount of solar radiation that is reflected back into space [see, for example, the atmospheric model described by K. Taylor and J. Pinner, Response of the climate system to atmospheric aerosols and greenhouse gases, *Nature*, **369**, 734–737 (1994)].

33. J. H. Martin and co-workers [Testing the iron hypothesis in ecosystems of the equatorial Pacific Ocean, *Nature*, **371**, 123–129 (1994)] seeded a 64-square-km area of the open Pacific with 15,600 liters of iron solution, and found a dramatic increase in the biomass of phytoplankton. It is the increased photosynthetic activity of the enhanced biomass that absorbs the atmospheric CO_2.

34. E. Teller, L. Wood, and R. Hyde [Global warming and ice ages: I Prospects for physics-based modulation of global change. 22nd International Seminar on planetary emergencies, Erice (Sicily), Italy (1997)] argue that the predicted increase in global warming due to the enhanced CO_2 abundance can be offset by a 1 percent reduction in the amount of solar radiation reaching Earth. They also argue that the onset of ice ages could be canceled by the appropriate manipulation of atmospheric CO_2 concentrations and the amount of solar energy reaching Earth.

35. James Peason and co-workers [Earth rings for planetary environment control, 53rd *International Astronautical Congress, 2002 (IAF-02-U.1.01)*] modeled various ring systems extending out to distances between 1.2 and 1.8 Earth radii. The ring system casts a variable width and location shadow upon Earth's surface, and it is this shading effect that results in atmospheric cooling.

36. O'Leary and co-workers [Retrieval of Asteroidal Materials, Space Resources and Space Settlements, *NASA SP-428*, 173–189 (1979)] studied the methods by which material from the Moon and Earth-approaching asteroids might be mined for chemical processing and the fabrication of large space structures. Indeed, O'Leary et al., conclude in their report that "immediate studies on asteroidal-retrieval mission opportunities" be initiated, with "technology readiness for asteroid retrieval by the mid-1980s". Twenty years late perhaps, in 2005, the first direct sampling of an asteroid surface was completed (hopefully) by the Japanese Space Agency's *Hyabusa* spacecraft. The term 'hopefully' is employed since it is currently unclear if material was actually collected. We will only know for sure when the spacecraft returns to Earth in June 2010.

37. David Keith [Geoengineering, *Nature*, **409**, 420 (2001)] discusses how the damaging side effects of pollution, derived from human industrial activity, might be redressed by planetary scale environmental engineering. He concludes: "I judge it likely that this century will see serious debate about—and perhaps implementation of—deliberate planetary-scale engineering."

38. O'Neill outlines his plans in The Colonization of Space [*Physics Today*, September (1974)]. His study concludes that, given the

appropriate funding and political commitment, space colonies could be constructed "without robbing or harming anyone and without polluting anything." He also argues that virtually all industry could be moved away from Earth and into space within 100 years. Writing these notes 32 years from the time of O'Neill's original publication, it is clear that his ideas have not, as of yet, found the required imagination, funding and political will for action from either industry or national governments.

39. P. Birch, Supramundane planets, *J. Brit. Interplan.* Soc. **44**, 169–182 (1991). The name *supramundane* is derived from the Latin *supra* (above) and *mundus* (world). The shell of a supramundane planet is not physically anchored to the surface of its central body and, consequently, the Jovian planets could be utilized as regions of colonization. A suprajupiter shell would have a surface area some 316 times that of Earth and potentially house, feed, and support up to 300 billion people.

40. D. R. Criswell [Solar system industrialization: implications for interstellar migrations, *Interstellar Migration and the Human Experience*, R. Finney and E. Jones (eds.) University of California Press, Berkeley (1985). pp. 50–87] has described human industrial development in terms of a triangular pyramid. The base of the pyramid is labeled according to 'skill,' 'matter,' and 'energy.' As human civilization has learned to control and use these three 'bases,' so it has produced the tip of the pyramid that Criswell calls "cumulative controlled connectives," or CCCs. It is the future growth of CCCs that will determine human destiny and, increasingly, the 'matter' and 'energy' bases markers of the industrial growth pyramid that will 'demand' the extraction and exploitation of the raw materials residing within the greater Solar System beyond Earth.

41. M. J. Fogg [Solar exchange as a measure of ensuring the long term habitability of Earth, *Speculations in Science and Technology*. **12** (2), 153–157 (1988)] builds upon the idea of star lifting described by Criswell (see Note 40 above and the detailed discussion in Chapter 5).

42. Philosopher Nick Boston [Astronomical waste: The opportunity cost of delayed technological development. *Utilitas*, **15**, 308–318 (2003)] takes this argument even further and argues that we (indeed, any galactic civilization) should strive to embrace and utilize the entire gamut of resources available within the Milky Way galaxy (and the Virgo super cluster beyond). That is, he argues humanity should strive to become a Kardashev Type III civilization (see Chapter 1, and especially Note 18 in that chapter). Indeed, he opens his paper in the following way: "As I write these words, suns are illuminating

and heating empty rooms, unused energy is being flushed down black holes, and our great common endowment of negentropy is being irreversibly degraded into entropy on a cosmic scale. These are resources that an advanced civilization could have used to create value structures, such as sentient beings living worthwhile lives."

43. The idea of a dynamic relationship between human population levels and the land carrying capacity was first articulated by Thomas Malthus (1766–1834) in 1798. The Malthusian catastrophe is often presented in the form that it is only such factors as epidemics and wars that hold human population levels in check. That is, if left unchecked the population level would simply increase to a point whereby the demand for food would outstrip all possible supply. Presumably (or at least hopefully) our descendants will solve the problem of epidemics and become more peaceful. If one follows the arguments presented by Criswell (Reference 40), then a Malthusian meltdown can be avoided by taking agriculture into space, indeed by adopting, for example, the space colonization strategies suggested by O'Neill (see Note 38).

3. The Sun, Inside and Out

The Sun is not a typical star for at least three reasons. First, it is a G2 spectral type star.[1] Second, it is not part of a multiple-star system, and third, we live on a planet that orbits around it. The fact that we orbit a star such as the Sun (Figure 3.1) is perhaps not especially surprising since, as discussed in Chapter 1, the Weak Anthropic Principle stipulates that the time for us to evolve T_{US} must satisfy the inequality $T_{US} = 4.5 \times 10^9 < T_{MS}$, the Sun's main-sequence lifetime. The greatest mass that our Sun might possibly have had would correspond to the condition $T_{MS} = T_{US}$, which implies a mass no greater than about 1.3 $M_\odot$. Any star with a mass larger than this limit will have $T_{MS} < T_{US}$. We will discuss the relationship between the parent star and the habitable planetary zone, where life can exist, in more detail in the next chapter.

The most common type of star is smaller, less massive, less hot at the surface, and of lower luminosity than the Sun. These are the M dwarf stars,[1] which constitute some 75 percent of all the stars within our galaxy. The G2 spectral-type stars, such as our Sun, make up about 8 percent of the entire stellar inventory. In addition, the available data on star companionship suggests that of order 80 percent of stars reside within multiple systems so the Sun, being a single star, is in the minority group. The Sun is certainly unique in that it is the only known star to be supporting an evolving system of life forms. Of the other billions of Sun-like stars in the Milky Way galaxy, the number that might be harboring extraterrestrial life (intelligent or otherwise) is completely unknown (recall Drake's Equation in Chapter 1). We will pick up the discussion on extrasolar planetary systems[2] again in Chapter 4. For this chapter, however, we will look more closely at the internal workings of the Sun. What makes the Sun tick and why does it puff up into a luminous giant with increasing age? Indeed, by fully understanding the canonical structure of the Sun

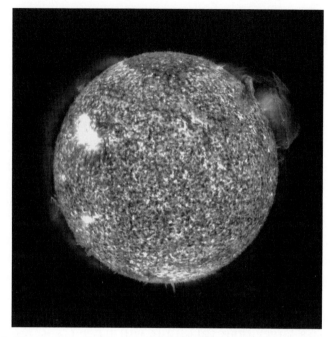

FIGURE 3.1. The Sun in ultraviolet light. This image from the SOHO satellite observatory shows a massive solar prominence (upper right) that has been produced by the lifting of hot plasma above the Sun's surface by twisted magnetic fields. (Image courtesy of ESA/NASA)

and how it changes with time, our solar-rejuvenating descendants can attempt to devise methods that will stop its gigantism in its tracks.

Star Basics

In a number of ways stars are very simple objects: they are massive, hot-gas spheres with a long-lasting internal energy source. This being said, there are many subtle complexities associated with the internal workings of a star and the manner in which its physical properties change with time. This chapter will not present the differential equations that describe stellar interiors nor describe in detail the physics associated with stellar evolution.[3] The intention here is to provide a straightforward order of magnitude overview of the physical properties of stars and tease out the main relationships between key quantities.

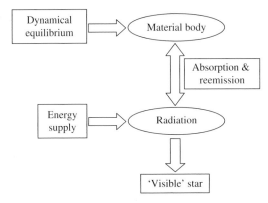

FIGURE 3.2. Schematic diagram of Eddington's two-component star model. The material body is made of electrons, ions, and atoms, and the radiative body is composed of electromagnetic radiation. Neither body would be stable for long if it were not linked to the other via atomic interactions.

In the book *The Internal Constitution of the Stars*, published in 1926, Arthur Eddington[4] described a star as being two bodies, superimposed (Figure 3.2). In this manner a star may be thought of as a material body (the atoms, protons, neutrons, and electrons) that make up its physical mass, and a radiative body composed of 'ethereal' photons (electromagnetic radiation). The two bodies – the material and the radiative – are symbiotic structures, and one without the other would soon become something that is not a star. The material body must be stable enough to prevent gravitational collapse, and the radiative body must be constrained so that its constituent photons do not leak out of the material body too quickly. For example, the time for a photon to travel unimpeded from the center of the Sun to its surface is $T_{UP} = R_\odot/c = 3.2$ seconds. Yet, as we shall explain below, the actual photon dwell time within the Sun is of order 5×10^{12} times larger than T_{UP} and this, it turns out, is essential if the material body is to remain stable against gravitational collapse.

The Dynamical Timescale

A fundamental observation concerning the Sun is that it is neither significantly expanding nor significantly contracting with time. This simple observation tells us that the Sun must be stable against gravitational collapse and that it has a long-lasting internal energy supply. To better understand this situation, we can ask what the dynamic collapse time (T_{DYN}) of the Sun would be if all of the forces opposing gravity suddenly vanished. To answer this consider a small mass m_{blob} of material at the Sun's surface (Figure 3.3). The gravitational interaction between this blob of material and the rest of the Sun will be $F_{Grav} = G\, M_\odot\, m_{blob}/R_\odot^2$, where G is the universal gravitational constant. Now, Newton's second law of motion tells us that this force will cause the blob of material to accelerate towards the Sun's center such that $F = m_{blob}\, a_{blob} = F_{Grav}$, where a_{blob} is the acceleration. Since the characteristic time of the collapse is T_{DYN} we can describe the acceleration as $a_{blob} \approx$ (change in speed/T_{DYN}) $\approx$ (change in distance/T_{DYN})/T_{DYN} $= R_\odot/(T_{DYN})^2$. Substituting this expression for a_{blob} into Newton's second law formula reveals the characteristic collapse time:

$$T_{DYN} = [R_\odot^3/GM_\odot]^{1/2} \qquad (3.1)$$

which, upon substitution for the Sun's radius and mass reveals T_{DYN} = 26.6 minutes—yes, minutes! If the material body of the Sun were not held in check by forces opposing gravity it would collapse into a black hole in about half an hour. Clearly this has

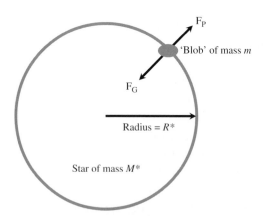

FIGURE 3.3. The dynamic collapse time is a measure of how long it would take an isolated blob of gas at the Sun's surface to travel to the center under unimpeded gravitational collapse. The unimpeded collapse of the Sun would produce a black hole with an event horizon some 6-km across in about half an hour.

not come about in the 4.56 billion years since the Sun formed, and the question now is why.

The short dynamical time does not mean that a star cannot change with time; it simply asserts a minimum time for any change. One can think of dynamical time as the sound crossing time of a star—this being the time required to 'communicate' across the star's diameter that a change has actually taken place.

Hydrostatic Equilibrium

The size of a star is determined by a number of competing forces. The gravitational force F_{Grav} continually works towards making a star as small as possible, while the outward pressure of the star's hot gas interior (F_{hot}) works toward making it as large as possible. The dynamical balance condition that determines the size of a star requires that at all points within its interior $F_{total} = F_{Grav} + F_{hot} = 0$. The balance condition comes about because if $F_{total} > 0$, then the star would expand, while if $F_{total} < 0$, the star would contract. This condition of hydrostatic equilibrium is very useful and indicates that the pressure and temperature of a star must increase inward towards the center.

To see why this is so, let's first consider the variation in pressure. The surface gas pressure (P_S) must, by necessity, be zero, since beyond the star's surface there is no more gas.[5] To estimate the central pressure (P_C) consider the star to be split into two hemispheres (Figure 3.4), each of mass $M/2$, where M is the total mass of the star under consideration. The gravitational centers of these two hemispheres will be a distance R apart, where R is the radius of the star, and the area over which the two hemispheres 'interact' is $A = \pi R^2$. Pressure is defined as being the force acting per unit area. The force holding our split star together is the gravitational attraction between the two hemispheres $F_{Grav} = G (M/2) (M/2)/R^2$, and the area over which this force is spread is A. Hence, the central pressure will be approximately:

$$P_C \approx F_{Grav}/A = [G/4\pi]M^2/R^4 \qquad (3.2)$$

For the Sun, Equation (3.2) indicates $P_C \sim 9 \times 10^{13}$ Pascal. (This value of P_C is actually about a factor of six too big when compared

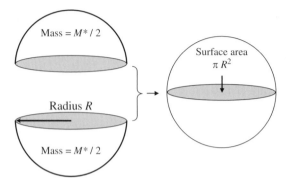

FIGURE 3.4. To estimate the central pressure of a star, consider two hemispheres, each containing half the mass of the star, with centers a distance R apart interacting over a surface area of πR^2.

with the results from detailed numerical calculations.) So, we now have our result: the pressure inside of a star increases inward, from zero at the surface to $\sim 10^{14} \, (M/M_\odot)^2/(R/R_\odot)^4$ Pa at the center.

That the pressure increases towards the center of a star makes sense, since the hydrostatic equilibrium condition requires that successive regions within a star must support the weight of overlying layers. As the entire weight of the overlying star must be supported by the central region, it makes sense that the pressure there must be highest.

The Pressure Law

The gas in the interior of a star provides the pressure that supports the weight of overlying layers because it is a hot gas. Indeed, the ideal gas equation links the pressure P at any point in the star to the local temperature T, the density ρ, and the number of particles per unit volume n. The relationship is $P = n \, kT$, where k is the Boltzmann constant.[6]

To determine the number of particles per unit volume, we must first describe the basic particle makeup of stellar material. Astronomers usually express the composition of a star in terms of the mass fractions of hydrogen (X), helium (Y), and all the other elements lumped together (Z). Astronomers also call the elements other than hydrogen and helium metals or heavy elements. By definition X + Y + Z = 1 and, for example, if 74 percent of the mass of a star is in the form of hydrogen, while 25 percent is in the form of helium, then 1 percent must be in the form of all other elements, and X = 0.74, Y = 0.25, and Z = 0.01. Now, given the high temperatures that prevail in stellar interiors the various atoms will be fully ionized, with all their electrons stripped away. In this manner, the stellar gas can be thought of as a 'soup' of positively charged ions and freely moving electrons. Table 3.1 indicates the number of ions and electrons that will be produced when the stellar material is fully ionized.

From Table 3.1 it can be seen that each ionized hydrogen atom will contribute one electron and one ion to the number density. Each ionized helium atom, on the other hand, will contribute three particles, two electrons and one nucleus, to the number density. The ion contribution from the heavy elements is going to be very small, first because Z itself is small, and second because the atomic mass number[7] A is generally large. The next element in the Periodic Table after helium is lithium, and this has an atomic number of A = 7. Consequently, this term is usually ignored in the final sum for the particle number density. The number of electrons contributed by the ionized heavy elements will typically be half the atomic mass (i.e., <A>/2). Bringing together the various terms described in Table 3.1, the number density of particles in a fully ionized gas is:

$$n = \frac{\rho}{m_H}\left(2X + \frac{3}{4}Y + \frac{Z}{2}\right) = \frac{\rho}{\mu\, m_H} \qquad (3.4)$$

The μ term introduced in Equation (3.4) is called the *mean molecular weight*, and the ideal gas equation corresponding becomes: $P = (k/\mu m_H)\rho T$. At this stage we note that, for a star composed entirely of hydrogen, X = 1 and μ =½. The initial

Table 3.1. The number of ions and electrons produced as a result of complete ionization. The <A> term corresponds to the average value of the atomic mass number[7] of the heavy elements, ρ is the density of stellar material, and m_H is the mass of the hydrogen atom.

	Number of ions	Number of electrons
Hydrogen	X (ρ/m_H)	X (ρ/m_H)
Helium	Y $(\rho/4m_H)$	Y $(\rho/4m_H)2$
Metals	Z $(\rho/<A>\, m_H) \approx 0$	Z $(\rho/<A>m_H)\,(<A>/2) = Z\,(\rho/2m_H)$

composition of the Sun, as determined by the chemical analysis of meteorites,[8] is believed to be X = 0.71, Y = 0.27, and Z = 0.02, with a corresponding mean molecular weight of μ = 0.613.

The Central Temperature

The ideal gas equation indicates that the pressure will increase provided the product ρT increases, and since the pressure required to support the weight of overlying layers increases with increasing depth, the temperature of a star must also increase inward towards the center. Again, to first order approximation, let us consider a star of mass M and radius R made entirely of hydrogen. Let us also assume a uniform density model. In this case, $\rho = M/[(4/3)\pi R^3]$. Using this value of the density with the ideal gas equation and substituting for P_C in Equation (3.2), an expression for the central temperature T_C can be derived:

$$T_C = [G\mu m_H/3k]M/R. \qquad (3.5)$$

For the Sun, Equation (3.5) indicates a central temperature of $T_C \approx$ 5 x 10^6 K. This value is about a factor of three on the low side compared to more detailed model calculations that find $T_C \approx$ 15 x 10^6 K.

We now have a series of expressions that characterize the properties of the material body of a star. Next we need to consider the radiative body, and then we need to identify the ways in which the material and the radiative bodies (recall Figure 3.2) interact.

Photon Diffusion Time

The temperature at the center of the Sun is some 15 million degrees Kelvin. At such high temperatures the characteristic wavelength of electromagnetic radiation is $\lambda \approx$ 2 x 10^{-10} meters, well inside the X-ray to γ-ray region of the electromagnetic spectrum. The radiation emitted at the surface of the Sun (where T_S = 5780 K), however, has a characteristic wavelength of $\lambda \approx$ 5 x 10^{-7} meters, in the visible part of the electromagnetic spectrum. Clearly, something has happened to the radiation during its journey from the Sun's center to its surface. Indeed, rather than taking an unimpeded straight-line path from the center to the surface (in a travel time of T_{UP}), the photon zigzags along a random walk path. The typical distance l traveled by a photon between interactions with a particle (the so-called mean free path length) amounts to $l \sim 0.1$ to 1.0-mm throughout most of the Sun's interior. The mean free path is actually related to the density of stellar material ρ and the opacity κ (which is a measure of how much the stellar material hinders the passage of radiation through it) by the relationship $l = 1/(\kappa\rho)$. The opacity is a complex term to quantify; it basically depends, however, upon the temperature, density, and composition of the stellar material. There is no simple or universal formula that describes the opacity of stellar material. (There are, however, a few approximations that apply under specific circumstances, as described below.) This being said, we need not worry about the details here. Indeed, the timescale of the random walk process by which the photons generated at the center of a star diffuse outward can be well approximated by a simple fixed-step random-walk model (Figure 3.5).

In a random-walk process the direction of motion changes abruptly after the photon has traveled a distance l, and the radial displacement from the center D after n steps will be approximately $D^2 = n\, l^2$. So the typical number of random-walk steps experienced by a photon as it moves from the center to the surface of a star is $n = (R/l)^2$. Here we have simply set the radial displacement $D = R$, the star's radius. Each step in the random-walk process takes the photon a time $T_{step} = l/c$ to accomplish, where c is the speed of light, and consequently the photon diffusion time T_{PD} is:

$$T_{PD} = nT_{step} = R^2/(lc) \qquad (3.6)$$

In passing from the center of the Sun to its surface, each photon typically undergoes $n = 5 \times 10^{25}$ interactions, taking of order $T_{PD} = 5 \times 10^5$ years to complete the journey.

It is the temperature gradient within a star – $\Delta T/\Delta R$ – that allows the energy generated at the center to flow outward. Characteristically, for the Sun, $\Delta T/\Delta R \approx (T_C - T_S)/R_\odot = 0.02$ K/m. That is, the temperature drops by two one-hundredths of a degree for

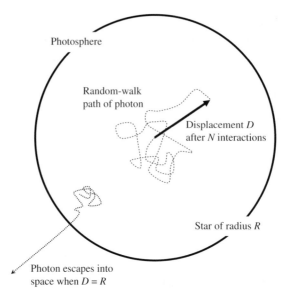

FIGURE 3.5. The random-walk path of a photon moving from the center of a star outward. After N random direction steps the radial displacement from the center is D. Once the radial displacement D of the photon is equal to the radius of the star R, it will escape into space, carrying away energy. The region where photons escape into space is called the *photosphere*.

every meter moved outward towards its surface. This is a small temperature gradient within the Sun, but it is enough to allow the energy to flow from the center to the surface, where it is lost into space. The important point at this stage, however, is that since the photons typically move less than a millimeter between interactions, they essentially 'sense' a radiation field corresponding to that of a blackbody radiator of temperature $T(r)$, where r is the distance from the center. This is why for the Sun the characteristic wavelength of the radiation emitted into space corresponds to that of a blackbody radiator of temperature $T_S = 5780$ K, rather than one of $T_C = 15$ million K.

Energy Transport

There are three possible modes by which energy can be transported: conduction, convection, and radiation. The first of these modes, conduction, occurs when energy is transferred by direct collisions and is correspondingly important within evolved stars that have regions of very high density (i.e., white dwarfs). The convective mode of energy transport requires the mass motion of a hot gas, while radiative transport involves the steady (albeit via a random-walk) outward flow of photons. It is the local temperature gradient at each point within a star that actually determines the mode by which the energy is transported outward. When the temperature varies rapidly with distance (that is, when the temperature gradient is steep), the energy will be transported by convection; otherwise, the energy transport is via radiation. In the case of the Sun the so-called convective stability condition dictates that energy transport is by convection over the outer third of its radius.[9] Within the inner two-thirds of the Sun (by radius) the energy transport is via radiation.

The sizes of the convective and radiative zones within a star vary according to its age and its initial mass. Although the Sun has a radiative core and a convective envelope, stars with masses greater than about 1.2 $M_\odot$ have convective cores and radiative envelopes (see Figure 3.6); stars less than about 0.3 $M_\odot$ in mass have fully convective interiors. The reason for the changeover in

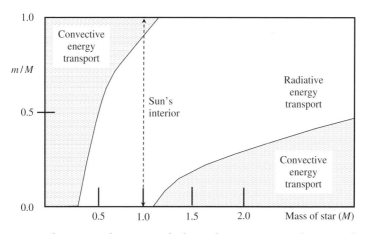

FIGURE 3.6. Schematic diagram of the relative size and type of energy transport zones within hydrogen-burning stars of different mass. The y-axis indicates the fractional mass, varying between the center at $m/M = 0$ to the surface at which $m/M = 1$. The shaded regions indicate those zones that are convective. Above a mass of 1.2 $M_\odot$ stars have convective cores and radiative envelopes; below 0.3 $M_\odot$ stars have fully convective interiors.

energy transport relates to the mode of energy generation itself and the opacity. For low-mass stars the internal temperatures are low enough [a result that follows from Equation (3.5)] that the gas is only partially ionized, which is a situation conducive to convective energy transport. Above 1.2 $M_\odot$ the so-called CN fusion reaction network switches on; this mode of energy generation is highly temperature sensitive, again a situation conducive to the onset of convective motion.

The Surface Temperature of a Star

If a star is in a steady state, neither appreciably heating up nor cooling down with time, the energy radiated into space at its surface must be exactly compensated for by the energy generated deep within its interior. We can express this condition in terms of the surface and central energy flux. At the surface the photon leaves the star at the speed of light c, and consequently the surface energy flux will be $(a\, T_S^4)\, c$, where a is the radiation density constant, and T_S is the star's surface temperature. Now, this surface energy flux must be matched by the energy flux at the star's center, which has moved through the star at an effective speed V_E determined by the photon diffusion process. That is, $(a\, T_S^4)\, c \approx (a\, T_C^4)\, V_E$, where T_C is the central temperature. This expression rearranges to give a relationship between the surface and central temperatures:

$$T_S = (V_E/c)^{1/4}\, T_C \tag{3.7}$$

The effective diffusion speed of the photon is given by $V_E = c/n$, where n in this case is taken to be the direct number of mean-free path lengths between the star's center and its surface, and since $n \gg 1$, Equation (3.7) tells us, as expected, that $T_S \ll T_C$. For the Sun we have with a direct walk $n = (R_\odot/l) \approx 6.9 \times 10^{11}$ (taking $l = 1$-mm) and $T_C = 4 \times 10^6$ K from Equation (3.5). Hence, we find a surface temperature of $T_S \approx 4400$ K. The Sun's actual surface temperature is 5,800 K.

Stellar Luminosity

The luminosity L of a star is defined as the total amount of electromagnetic energy radiated into space per second. Taking a star to be a spherical blackbody radiator of radius R and surface temperature T_S, the luminosity can be determined through the Stefan-Boltzmann law:

$$L = 4\pi R^2 \ \sigma \ T_S^{\ 4} \tag{3.8}$$

Now, if we substitute for T_S from Equation (3.7) and use the identity $\sigma = a\, c/4$, where c is the speed of light and a is the radiation density constant, so Equation (3.8) can be rearranged as:

$$L \approx [(aT_C^{\ 4})V]/[R^2/(lc)] = [\text{radiant energy content}]/T_{PD} \tag{3.9}$$

where $V = (4/3) \ \pi \ R^3$ is the volume of the star. Equation (3.9) tells us, therefore, that the luminosity of a star is governed by the rate at which the radiant energy content of the star is lost into space via photon diffusion.[10] For a given amount of energy within a star, Equation (3.9) indicates that the longer the photon diffusion time, the smaller the luminosity. Equation (3.6) further tells us that the photon diffusion time increases as the mean-free path l between photon interactions with stellar material decreases. In other words, if the opacity κ of the stellar material increases (so that l decreases), the luminosity of the star will also decrease.

Although the mass and radius of a star's material body determines its central temperature T_C (as indicated by Equation3.5), it is the mean-free path of the photons that determines the star's surface temperature and luminosity. If the mean-free path length of the photons within a star is reduced (by, for example, increasing the opacity κ), then the luminosity will also be reduced as a result of the increased photon diffusion time (thanks to Equation 3.9). Likewise, the surface temperature will be reduced because of the reduction in the effective photon diffusion speed V_E (as described by Equation 3.7).

Energy Generation

The Sun radiates energy into space at a rate of $L = 3.85 \times 10^{26}$ joules per second (or watts), and yet it does not cool off with time (i.e., T_S is constant on timescales well in excess of T_{PD}). This indicates that the Sun must have an internal energy source. Further, since on a timescale of at least several T_{PD} the Sun's radius is neither increasing nor decreasing, it is correspondingly not suffering any net loss or net gain of internal energy. That is, there must be an exact balance between the energy lost and radiated into space at the surface of the Sun, and the energy generated deep within its interior.

A star can tap two main energy sources during its lifetime. It can contract and thereby feed off the gravitational potential energy liberated, or it can feed from the energy liberated by nuclear fusion. While feeding on its gravitational energy a star will become physically smaller and hotter at its center, but its composition will remain unchanged. If it generates internal energy via nuclear fusion reactions, however, then it will maintain a near constant radius and central temperature, but its internal composition will change with time. A star can, in effect, turn on and turn off the gravitational potential energy source as required. That is, if the nuclear fusion reactions are unable to provide the energy to power a star, it will begin to contract on the so-called Kelvin-Helmholtz timescale (T_{KH}) which is given by the ratio of the gravitational potential energy of the star divided by its luminosity (i.e., its rate of energy loss). Correspondingly, for stars of mass M, radius R and luminosity L,

$$T_{KH} \approx [GM^2/R]/L \qquad (3.10)$$

where G is the universal gravitational constant. For the Sun, $T_{KH} \approx 2 \times 10^7$ years, and the contraction rate would amount to about 75 meters per year if the Sun were powered entirely by contraction. The release of gravitational potential energy not only results in the star becoming smaller; it also causes the central temperature to increase, as indicated by Equation 3.5 when R is reduced for a fixed mass M. This effect is important since, for example, at the end of a star's main-sequence phase (i.e., with the exhaustion of hydrogen in its core; see below), it is the contraction of the central regions that causes the central temperature and density to increase, a situation that continues until the triple-alpha reactions—during which helium is converted into carbon—can commence.

Nuclear Fusion

Stars such as the Sun generate energy within their central regions by the nuclear fusion reactions that transform four hydrogen nuclei (4 protons) into a helium nucleus (containing 2 protons and 2 neutrons). Schematically, $4H \Rightarrow {}^4He$ + energy (see Figure 3.7). Such fusion reactions will run efficiently once the temperature exceeds $T_{fusion} \sim 10$ million K. The region over which this temperature condition is met within the Sun can be determined from the central temperature and the average temperature gradient. Specifically, it is required that $T_C - [\Delta T/\Delta R]r > T_{fusion}$, which indicates that $r < (T_C - T_{fusion})/[\Delta T/\Delta R] = 5 \times 10^6/0.02 \sim 2.5 \times 10^8$ meters $\sim R_{\odot}/3$. Hence, the energy radiated at the surface of the Sun is entirely generated within the inner third (by radius) of its interior.

The conversion of hydrogen into helium proceeds in the Sun via the proton-proton (PP) chain,[11] and the details of the reaction network were described by Hans Bethe in the late 1930s. Specifically, Bethe realized that during the first step of the chain two things must happen. First, the two protons must approach one another so closely that there is a non-zero probability that they will overcome their mutual electrostatic repulsion. This condition is determined by the so-called Gamow factor (since the

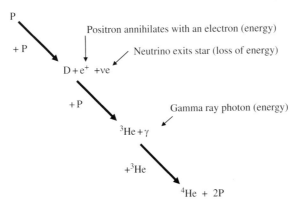

FIGURE 3.7. Schematic flow of the PP chain. The end result of the chain is that $4P \Rightarrow {}^4He + 2e^+ + 2\nu_e$ + energy. Here e^+ is a positron, the antimatter equivalent of the electron, while ν_e is an electron neutrino; both are generated during the inverse beta decay step. The energy liberated per conversion is just over 4×10^{-12} joules.

details of quantum mechanical 'tunneling' were first determined by George Gamow in 1928). Second, at the same time that the protons undergo this very close approach, one proton must undergo an inverse beta decay to produce a neutron. In this manner a deuterium nucleus, consisting of a bound proton and neutron, can be formed. In fact, stars only exist for long periods of time because the ^{2}He nucleus, consisting of two protons, is unstable and because the inverse beta decay occurs only rarely to produce deuterium. If it were not for the inverse beta decay requirement dramatically slowing down the initial P+P reaction, the Sun would have converted all of its central H into ^{4}He in a matter of a few hundred thousand years.[12] The average time interval required for a single proton to fuse with another proton, to produce deuterium, is of order 10 billion years. This time is, in fact, the nuclear (or main-sequence) timescale, and it is the characteristic time after which all the protons in the energy generation region of a star will have fused with other protons. In any given second, however, of order 10^{38} protons successfully undergo fusion reactions within the Sun, and the energy generated by these reactions will be radiated into space after a time T_{PD}.

The rate at which nuclear energy is generated per unit mass per second $\varepsilon(r)$ at radius position r within a star is determined by the local composition, temperature, and density: $\varepsilon(r) = \varepsilon\{composition, T(r), \rho(r)\}$. The full details of the energy generation calculation need not be followed here, but suffice it to say that a power law representation can be applied such that:

$$\varepsilon(r) = \varepsilon_0\, \rho\, X^2\, T^\alpha \tag{3.11}$$

where X is the hydrogen mass fraction of the star, $\rho(r)$ is the density, $T(r)$ is the temperature, and ε_0 and α are constants. For stars such as the Sun, where $T_C \approx 15$ million Kelvin, the exponent α has a value of about 4. The energy generation rate is clearly sensitive to changes in temperature, due to the size of the exponent α, and any process that reduces the central temperature and density will result in a decrease in the energy generation. Likewise, as one would expect, as the hydrogen in the core is gradually converted into helium, the central value of X will decrease and, ultimately, when X(core) = 0, the hydrogen fusion reactions will stop altogether.

The Mass Luminosity Law

As indicated above, the temperature gradient in the Sun is of order $T_C/R_\odot \sim 0.02$ K/m, and a photon typically travels less than a millimeter between interactions with the surrounding stellar material. In this manner we can treat the radiation field at radius r within a star as that due to a blackbody radiator of temperature $T(r)$. So, let us consider the flow of radiation through a thin shell of thickness Δr. At the base of the shell (at radius r), the Stefan-Boltzmann law tells us that the radiative flux $F(r) = \sigma T^4$ (watts/m^2). At the top of the shell the temperature is $T(r + \Delta r)$, and the flux will be F(r) $+ \Delta F = \sigma(T + \Delta T)^4 \approx \sigma(T^4 + 4T^3\,\Delta T)$. Here we are ignoring terms of $(\Delta T)^2$ and higher powers, as they will be very small. Now, ΔT is negative because the temperature must decrease outward from the center of the star, so the energy flux absorbed within the shell must be $\Delta F = 4\,\sigma T^3\,\Delta T$. This absorption of energy is related to the opacity $\kappa(r)$ of stellar material and, by definition, the energy flux absorbed across a region of width Δr will be $\Delta F = -\,\kappa(r)\,\rho(r)\,F(r)\,\Delta r$. Now, again by definition, the energy flowing across a shell of radius r each second (the luminosity) is $L(r) = 4\pi\,r^2\,F(r)$, so equating our two terms for ΔF we have: L(r) $= -\,(4\,\pi\,r^2)\,4\,\sigma T^3\,[\Delta T/\Delta r]/\kappa(r)$ $\rho(r)$. To make headway now, we need to express the opacity $\kappa(r)$ in terms of the composition, density, and temperature at radius r. For Sun-like and lower mass stars the opacity can be expressed in terms of the so-called Kramer's power law with $\kappa(r) = \kappa_0\,(1 + X)\rho$ $T^{-3.5}$, where κ_0 is a constant. For higher mass, higher temperature stars the opacity switches to that of electron scattering, which varies as $\kappa_{es} = 0.02(1 + X)$ m^2/kg, independent of the temperature and density.

At this stage we will combine the expressions for the luminosity, opacity, and the temperature gradient to determine how the luminosity varies with stellar mass M, the hydrogen mass fraction X, and the mean molecular weight μ. When the Kramer's opacity law holds true, the luminosity varies as the mass to the fifth power:

$$L = L_{KR}\frac{\mu^{7.5}}{(1+X)}M^5 \qquad (3.12)$$

where L_{KR} is a constant[13]. Equation (3.11) is the mass-luminosity relationship that approximately holds true for stars in the mass range $0.5 < M/M_\odot < 2.0$. For higher mass stars, where the opacity is mostly due to electron scattering, the luminosity varies according to the mass cubed. The data derived from binary star observations indicates that stellar luminosity varies as the approximate fourth power of the star mass ($L \sim M^4$), which indicates that the simplified arguments being presented in this chapter are, in fact, dimensionally correct and do indeed offer a reasonable description of stellar characteristics.

The key physical feature missing from the models described so far is that for the mode of energy transport. We have assumed that all the energy is carried by radiation, whereas detailed computer models indicate that stars can have extensive convection zones, and this will have an important effect upon their internal structure and the exact form of the mass-luminosity relationship. For stellar masses smaller than about 0.5 $M_\odot$, more than 50 percent of the star's outer envelope by mass is convective; for masses greater than about 20 $M_\odot$ more than 50 percent of the star's interior, again by mass, is convective.[14] All the above being said, the main point that will be of interest to future star engineers is that the luminosity of a star can be reduced by lowering its mass.

A Journey Through the HR Diagram

The Hertzsprung-Russell (HR) diagram is the historical battleground between theory and observation. The diagram displays the luminosity and temperature relationship of stars, and it reveals that important correlations exist between the two quantities (Figure 3.8). Of prime importance is the delineation of the main sequence that stretches along the diagonal from the hot, high luminosity (high-mass) stars to the low luminosity, cool (low-mass) stars. Indeed, over 90 percent of the observed stars fall on the main-sequence diagonal in the HR diagram. Those stars not on the main-sequence fall in either the red giant or the white dwarf regions.

Detailed computer models have shown that the main-sequence is delineated by those stars that are generating internal

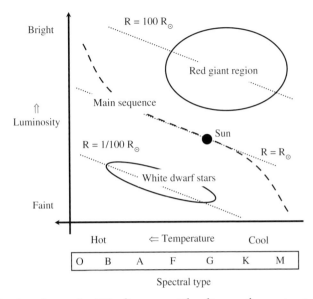

FIGURE 3.8. A schematic HR diagram. The lines of constant radius can be placed in the diagram according to the Stefan-Boltzmann relationship expressed in Equation (3.8).

energy through the conversion of hydrogen into helium. The red giant region, on the other hand, is populated by those stars that are generating internal energy through the conversion of helium into carbon.[15] Further, the white dwarf region is populated by old, low-mass stars that are in fact simply cooling off,[16] their days of producing internal energy through fusion reactions having ended. These corpse stars, given enough time, will eventually become zero luminosity, zero temperature black dwarfs.

Detailed computer models also indicate that the manner in which the surface temperature and luminosity of a star vary with time is dependent upon its initial mass. Stars end their formation stage by initiating hydrogen fusion reactions and correspondingly settle onto the main sequence in the HR diagram. All stars[17] go through a hydrogen fusion stage on the main sequence and a helium fusion stage in the red giant region (Figure 3.9). Stars more massive that 8 $M_\odot$ can initiate fusion reactions beyond that of the triple alpha reaction, but they eventually end their days as a supernova—literally blowing themselves apart after the formation of an iron-rich central core. A neutron star[18] remnant may or may

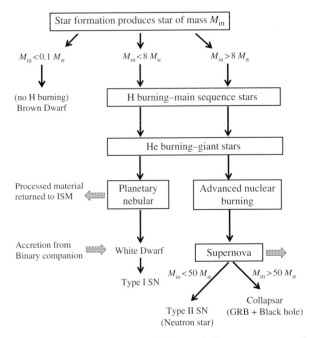

FIGURE 3.9. Evolutionary pathways followed by stars according to their initial mass. The horizontal block arrows indicate stages where mass is either lost into space or accreted by the star if it chances to be in a binary system.

not be produced during the rapid supernova phase; astronomers are still debating the exact details. Stars with an initial mass of less than 8 $M_\odot$ are unable to initiate fusion reactions beyond that of helium burning and consequently become white dwarfs after undergoing a visually dramatic planetary nebular stage.

The Journey of the Canonical Sun

As the Sun ages it steadily consumes the hydrogen within its core—the region encompassing the inner third (by radius) of its interior. Indeed, it is the change in the core's composition that drives the star's evolution towards a hotter, more luminous, and larger configuration (Figure 3.10). As the hydrogen stored in the Sun's central core is depleted by PP fusion reactions, so $X \Rightarrow 0$ and μ increases from 0.613 (its zero-age main-sequence value) to 1.316 at core hydrogen exhaustion. Equation (3.12) indicates that this

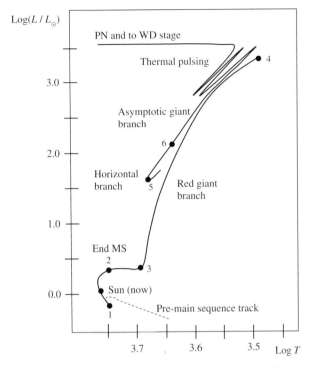

FIGURE 3.10. The Sun's journey through the HR diagram. The time intervals between the numbered points are given in Table 3.2.

composition change should result in the luminosity increasing by a factor of order $(1.316/0.613)^{7.5} \sim 300$. Such a dramatic increase in the Sun's luminosity during its main-sequence lifetime will not actually occur, because it is only the inner core that experiences the compositional change; the outer envelope maintains the original solar composition. It is the chemical discontinuity at the core boundary, however, that will eventually cause the Sun to become a red giant, and if nothing is done about it, it is at this stage that the Sun will cause all life on Earth to become extinct (as we will discuss in the next chapter).

We need not follow the detailed story of the Sun's formation here. Needless to say, however, it formed through the gravitational collapse of a low density, low temperature, and extended cloud of gas. The time for the Sun to reach Point 1 in Figure 3.10 (its so-called zero-age main-sequence position) is determined by the rate at which material is accreted at the center of the solar nebula. The various detailed models for the Sun's formation suggest a proto-Sun stage lasting about 10 million years. At Point 1 (Figure 3.10) the PP chain reactions begin in the core, and the Sun becomes a *bona fide* star. Table 3.2 indicates that on the zero-age main-sequence the Sun is actually slightly less luminous and slightly smaller than is currently observed.[19] After 4.5 billion years the Sun, as we currently see it, is about middle-aged, with half of the hydrogen within its central core having been consumed. At Point 2 in Figure 3.10, the hydrogen within the core has all been consumed. This point is reached some 11 billion years (see Table 3.2) after the PP reactions first started. After central hydrogen exhaustion, the Sun begins moving rightward in the HR diagram, becoming cooler and larger. At Point 3 the luminosity and radius begin to increase rapidly, and the Sun starts to ascend the red giant branch. Deep in the interior of the Sun the temperature and density of the central core are now increasing rapidly, with energy being generated in a thin hydrogen 'burning' shell above the dormant core. Eventually, the temperature in the central core becomes hot enough (about 100 million degrees Kelvin) for helium fusion reactions[15] to begin. Indeed, the onset of helium burning is a veritable hot, explosive flash.

At the peak of the helium flash the Sun will find itself at the tip of the red-giant branch (Point 4), where it will be some 2,349 times more luminous than at present, and some 166 times larger (see Table 3.2). At this stage the planet Mercury will be destroyed,

Table 3.2. Characteristics of the solar evolutionary track shown in Figure 3.10. The ages given in the second column are in units of billions of years. The luminosity and radius values given in the third and fifth columns are expressed in units of the Sun's current luminosity and radius. (Table data based upon the model calculations by Sackmann, Boothroyd, and Kraemer; see Reference 20).

Stage	Time (t_9)	$L/L_\odot$	T (K)	$R/R_\odot$
1	0.0	0.70	5586	0.897
Now	4.5	1.00	5779	1.00
2	10.91	2.21	6517	1.58
3	11.64	2.73	4902	9.5
4	12.233	2349	3107	165.8
5	12.234	41	4724	9.5
6	12.345	130	4375	20

consumed within the Sun's bloated outer envelope. We will discuss this stage in greater detail in the next chapter. The Sun spends a very short amount of time at the red-giant tip (Point 4), perhaps a few hundred thousand years, eventually dropping substantially in luminosity to a position on the so-called horizontal branch, where it will begin the steady consumption of helium within its core (point 5). At this stage the Sun will be about 40 times more luminous than at present. The core helium burning phase is not as long lasting as the main-sequence phase, and within a hundred million years the Sun's central helium supply will become exhausted (Point 6). At this stage a number of complex internal processes are set up. The Sun now begins to ascend the asymptotic giant branch and hydrogen and helium fusion reactions are taking place within rings (or shells) around the carbon-rich central core. The hydrogen shell source is situated above the helium shell source, and the two 'furnaces' turn on and off in a complex series of interactions. At this stage the Sun undergoes what are called thermal pulses—periods of rapid expansion and contraction (in time intervals of several hundred days) accompanied by large swings in luminosity and temperature. Such stars are distinguished observationally as long-period Mira variables, named after the prototype system omicron Ceti (Mira).

Various numerical models have been developed by astronomers to describe the advanced thermal pulsing stage of the Sun.[20] However, these models do not, as of yet, offer a clear consensus on what might happen to Earth. Some models suggest the Sun will expand beyond 215 $R_\odot$, thus engulfing Earth and bringing its history to a final close. Other models suggest that the Sun won't expand quite so much and, consequently, Earth as a physical body will survive. Life, however, will have long perished because of the Sun's increased luminosity. The key unknown at this stage is exactly how much mass the Sun might lose during the red-giant and asymptotic giant phases. Present observations suggest that low-mass stars, such as the Sun, lose of order 0.1 to 0.2 solar masses in the form of a stellar wind during the post main-sequence phase. The numerical models including mass loss find that the Sun might not expand to the extent that Earth is engulfed during the thermal pulsing stage. Part of the reason why Earth, again as a physical object, survives when mass loss is included is because its orbital semi-major axis actually increases—a topic we shall look at again in the next chapter.

With the onset of thermal pulsing the Sun is beginning to truly die, and it is rapidly running out of fuel in those regions that are hot enough to power the hydrogen and helium shell sources. The Sun may be dying at this stage, but it will go out in a blaze of glory. The thermal pulsing results in the core and the envelope parting company, and over a period of several tens of thousands of years the outer envelope will be cast off into the surrounding interstellar medium. The ultraviolet photons produced by the hot, now-exposed carbon-rich core, however, begin to ionize the hydrogen at the innermost edge of the expanding gas envelope, and this leads to the formation of a planetary nebula. William Herschel, who first described such nebulae in the late 18th century, thought that such nebulae reminded him of planetary disks, and astronomers have continued to use his misnomer ever since.

The central, carbon-rich core evolves rapidly during the planetary nebula phase, and while initially very luminous ($L \sim$ 2500 $L_\odot$) and extremely hot (T $\sim$ 30,000 K), it soon enters the white dwarf region in the HR diagram (see Figure 3.8). As a white dwarf[16] the future Sun will gradually cool off and slowly fade out. Indeed, the cooling time of a white dwarf is immense. Since white dwarfs are not generating any energy within their interiors their total energy reserve is essentially the thermal energy of their constituent particles. To order of magnitude, the energy reserve that a white dwarf has to radiate into space is $E_{WD} \approx n\,k\,T$, where n is the number of particles, k is the Boltzmann constant, and T is a typical internal temperature. For a one solar mass white dwarf assumed to be composed entirely of carbon, $n \sim 10^{56}$; taking $T = 10^7$ K we then have $E_{WD} \sim 1.4 \times 10^{40}$ joules. Further, adopting a typical luminosity of $L_{WD} = 10^{-3}$ $L_\odot$, the cooling time will be $T_{cool} = E_{WD}/L_{WD} \sim 3.5 \times 10^{16}$ seconds $= 10^9$ yrs. The cooling time will actually be much greater than a billion years since as the white dwarf cools its luminosity decreases and consequently the thermal energy is radiated away less rapidly. Indeed, a more detailed calculation indicates that several tens of billions of years are required for a white dwarf to become a black dwarf.

The ultimate end state of the Sun will be that of a black dwarf, a cold object about the same size as Earth, with zero luminosity supported against gravitational collapse by its constituent degenerate electrons.[16] For the canonical Sun the deep future promises an infinitely[21]bleak outlook.

The Reasons for Gigantism

Before continuing with this chapter we should look at why stars become giants as they age. This being said, the exact physical reasons for why a star puffs up to become a giant when the central hydrogen supplies are depleted are not fully agreed upon by astronomers. Everyone agrees that it happens; indeed, it is an observational fact. But it appears that many subtle effects come into play in order to produce the tendency towards gigantism.

The various detailed computer models indicate that two distinctly different things begin to happen when all of the hydrogen within a star's core has been converted into helium. First, the core can no longer provide energy via PP fusion reactions to heat the stellar interior, and consequently gravitational contraction of the core (not the whole star) begins. The contraction timescale of the core will be that of the Kelvin-Helmholtz timescale described in Equation (3.10). As the core contracts it both heats up and acquires a higher density. This is good, since it will eventually result in the onset of helium fusion reactions.

At the same time the core begins to contract as a result of core hydrogen exhaustion, the outer envelope of the star begins to expand. In addition, hydrogen fusion reactions begin in a shell source surrounding the hydrogen-depleted core. It is generally argued that one of the principal reasons that the envelope expands at this stage is because of the composition difference between the core and the envelope. We can see that this should result in the star expanding by looking at the ideal gas equation. As described above, the pressure P is related to the density, composition (through the mean molecular weight μ), and the temperature T, such that $P = (k/\mu\, m_H)\,\rho\, T$. Across the core-envelope boundary,[22] the pressure and temperature must be constant; the density on the other hand must vary in step with the mean molecular weight, such that ρ/μ is constant across the boundary. Since the mean molecular weight varies from 0.6 in the core to 1.3 in the envelope, the density must correspondingly decrease by a factor of $0.6/1.3 \approx 0.5$ across the core-envelope boundary. This reduction in the density dictates that the star must expand in order to accommodate the amount of material situated above the core.[23] In addition to the core-envelope composition jump effect, experiments with numerical stellar models have shown that a star will swell up as a result of becoming more centrally condensed and as a result of having a hydrogen-burning shell situated above an electron degenerate core.[24]

The importance of the core-envelope composition jump for producing gigantism is further exemplified by the fact that stars less than about 0.25 $M_\odot$ in mass *do not* undergo a red-giant phase. The reason for this is that the interiors of these very low-mass stars are almost fully convective (recall Figure 3.6). The evolution of a 0.12 $M_\odot$ star is shown in Figure 3.11, and in contrast to the Sun's evolutionary track (Figure 3.10), this star evolves to higher rather

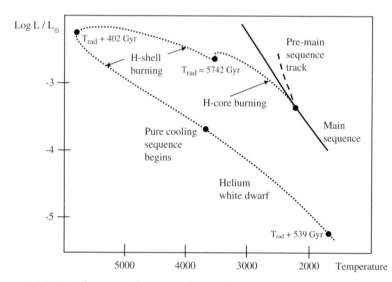

FIGURE 3.11. HR diagram showing the evolutionary path of a 0.12 $M_\odot$ star. The fully mixed interior of very low-mass stars (see Figure 3.6) prevents the core-envelope chemical discontinuity from coming about, and consequently these stars do not undergo a red-giant phase following hydrogen exhaustion. Figure based upon calculations by Gregory Laughlin and co-workers: *The Astrophysical Journal*, **482**, 420–432 (1997).

than lower temperatures with age, and there is no large change in the radius.

At time T_{rad} (see Figure 3.11) a small radiative core develops, and a small compositional gradient begins to grow with the continued depletion of hydrogen. The core, however, soon becomes inert, and a hydrogen-burning shell source develops. Eventually, all of the hydrogen that can possibly be consumed by the PP fusion reactions is used up (T_{rad} + 402 Gyr in Figure 3.11), and the star becomes a helium white dwarf, slowly cooling off into obscurity and reversing its direction of evolution in the HR diagram.

A Negative Feedback System

In this chapter we have attempted to provide an overview of the internal workings of stars. Perhaps the key point is that in general one can think of a star as a negative feedback system (see Figure 3.12) that is capable of finding an equilibrium size, surface temperature, and luminosity in accordance with its mass, its mode of energy generation, and its composition. If any one or all of the latter quantities are changed, then the equilibrium values for the temperature, size, and luminosity also change.

Figure 3.12 shows the essential arrangement of interconnections between surface and central stellar quantities. The line linking the central temperature to the opacity and the surface temperature in the figure corresponds to Equation (3.7). The relationships between the central and surface temperatures, the temperature gradient, and the radius establishes the feedback mechanism responsible for enablement of hydrostatic equilibrium, whereby the pressure gradient is capable of supporting the weight of overlying layers at each point inside of the star. The link between the central temperature and composition establishes the energy generation rate and luminosity as described by Equation (3.9) and Equation (3.11). The energy generated at the center of the star flows down the temperature gradient, from the hot center to the cooler surface, and is eventually radiated into space. The time for the radiation to traverse the temperature gradient (T_{PD}) is given by Equation (3.6).

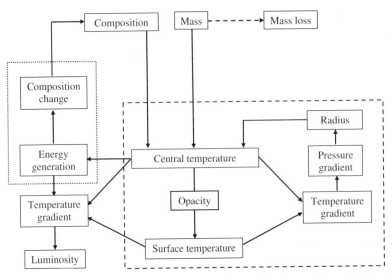

FIGURE 3.12. A schematic interaction chart showing the relationships between various stellar quantities. The dashed box in the lower right contains those quantities that are involved in establishing hydrostatic equilibrium. The box in the upper left contains those quantities that are responsible for driving stellar evolution. The dashed line indicating mass loss is also a mechanism for driving internal readjustment.

Once the mass and composition have been specified the radius, surface temperature, and luminosity of a star are established according to the operation of a stabilizing negative feedback mechanism. To see how this works, imagine that for some reason the central energy generation rate suddenly increases. Equation (3.11) tells us that such an increase must have come about because of an increase in the central temperature. Now an increase in the central temperature will cause an increase in the central pressure (thanks to the perfect gas law), and this in turn will result in an increase in the temperature and pressure gradients. An increase in the pressure gradient, however, will cause the star to expand. For a fixed stellar mass Equation (3.5) indicates that as the radius increases so the central temperature decreases and, hence, according to Equation (3.11), the energy generation rate will be reduced. In this negative feedback manner, an increase in the central temperature results in the star readjusting its internal structure such that the increase is damped out. Conversely, if the nuclear energy generation rate suddenly decreased, the response of the negative feedback mechanism would be to cause the star to shrink, thereby increasing the central temperature and consequently causing an increase in the energy generation rate. By the use of negative feedback mechanisms a star is able to remain stable against collapse (that is, in hydrostatic equilibrium), and it is able to generate exactly the right amount of energy in its central regions to compensate for the energy that it loses into space at its surface.

The reason why stars must evolve with time is illustrated in Figure 3.12. As a result of the generation of energy through nuclear fusion reactions (i.e., via the PP chain; see Figure 3.7), the composition of the central regions is changed with time. Specifically, the composition changes from one that is initially hydrogen-rich to one that is helium-rich and hydrogen-depleted. This change in the central composition will correspondingly result in a change in the mean molecular weight μ[given in Equation (3.4)] and the opacity κ of the stellar material. Because of the changes in these latter two quantities the star will have to find a new equilibrium temperature gradient resulting in a new luminosity, radius, and surface temperature. The mass-loss term shown in Figure 3.12 will also drive stellar change since, literally, the mass of the star is reduced over time—a point future asteroengineers will most definitely take note of.

Fundamental Constants

"From a drop of water a logician could infer
the possibility of an Atlantic or a Niagara without
having seen or heard of one or the other."

Arthur Conan Doyle

In this chapter we have been mostly concerned with describing what goes on in the interiors of stars. Now, however, we will address the issue of why stars have the actual characteristics that they do. It turns out, as you will see, that the observed properties of the stars are determined by the fundamental constants of physics.

We have already indicated (see Note 3 in Chapter 1) that the speed of light c is a fundamental constant. It is a limiting speed, and we can observe no object that travels faster than $c = 2.99792458 \times 10^8$ m/s in our universe. Other fundamental constants of our universe (in all their measured glory) are: $G = 6.6742 \times 10^{-11}$ m^3/kg/s^2, the universal gravitational constant; $h = 6.6260693 \times 10^{-34}$ Js, Planck's constant (fundamental to the quantum world); the proton mass $m_p = 1.67262171 \times 10^{-27}$ kg, and the unit of electrical charge $e = 1.60217653 \times 10^{-19}$ C. Both of the latter two constants are important in the description of atomic structure. If you change any one of these fundamental constants then you literally generate a universe with properties distinctly different from our own. There is, in fact, a rather restricted range of variation among the fundamental constants that will allow for the formation of stars and the existence of life. Other possible universes (sometimes called "world ensembles") might be entirely void of stars; still other universes might be full of very low-mass stars that are incapable of undergoing helium fusion reactions to produce the carbon atoms[25] essential for the emergence of life.

From the fundamental constants just given one can construct two dimensionless constants: $\alpha = e^2/\hbar c \approx 1/137$ and $\alpha_G = Gm_p^2/\hbar c \approx 5 \times 10^{-39}$ (here we have used the standard notation that $\hbar = h/2\pi$). These two constants – the electromagnetic fine scale constant and the gravitational fine scale constant, respectively – when combined with fundamental mass and length terms, appear to account for such observations as why the universe is as big as it is now, why stars are as massive as they are, and why atoms have their various properties.[26]

In the following discussion we are going to use a standard dodge, and rather than account for exact values, we will consider only the order of magnitude argument. That is, what is really important is the determination of whether a quantity is of order 10 in size, 100, 10^{-7}, 10^{23}, and so on. In this manner we will mostly be using the '~' sign rather than the strict '=' sign in the calculations that follow. This method, in which small constant factors such as 2 or π are ignored, is an example of the Fermi calculation approach discussed earlier (see Note 1 of Chapter 1). The idea is that by the end of a series of multiplications and divisions most of these small constant terms will cancel each other out to produce a number of order unity. This method usually works, but there are times when one has to be a little careful, so a dose of due diligence is also required.

The Quantum World of the Electron

To answer the question, "Why are there stars?' we must first take a brief excursion into the small-scale quantum world of the atom and, more specifically, look at how the electrons inside a star interact with one another. On the scale of the atom we must first abandon our everyday notions of what particles are. We are now in the quantum realm, where entities such as electrons, atomic nuclei, and photons have simultaneous wave-like and particle-like properties. The Nobel Prize-winning physicist Louis de Broglie (1892–1987) described this wave-particle duality by arguing that the wavelength λ of a particle is related to its momentum p through the relationship $p = h/\lambda$, where h is Planck's constant. Further, the energy of motion K of our particle is related to its momentum by the relationship $K = p^2/2m = h^2/2m\lambda^2$, where m is the particle mass. Where this becomes important to our story is that if a particle, say an electron, is confined to a region of size d, then its associated de Broglie wavelength must satisfy the condition $\lambda \leq d$. Since, however, a maximum wavelength corresponds to a minimum momentum, the energy of a confined entity (i.e., our electron) must be at least $K_0 \sim h^2/2 \, m_e \, d^2$, where $m_e \approx m_p/1000$ is the electron mass. In other words, an electron cannot be completely at rest even when it is confined. This behavior is important since it results in an outward pressure that tends to resist further confinement. Nobel Prize-winning physicist Wolfgang Pauli (1900–1958) introduced the quantum mechanical idea of the non-overcrowding of electrons[27] and, accordingly, if there are N electrons in a volume of space V with a characteristic dimension d, then the minimum energy of each electron is $K_0 \sim h^2/2 \, m_e \, d^2$, where $d^3 = (V/N)$. An electron gas constrained according to the Pauli exclusion principle is said to be degenerate. This effect is important for stars since the overcrowding pressure results in degenerate electrons being able to support the star against gravitational collapse.[28] We will come back to this point in a few moments, but first let's re-cast Equation (3.5), our expression for the central temperature of a star, in terms of the average inter-particle separation d.

Collapsing Gas Clouds

For a perfect gas,[6] Boyle's law provides a relationship between the pressure P, the temperature T, the volume V, and the total number of particles N in the gas, with $P V/T = N k$, where (again in all its derived glory) $k = 1.38065 \times 10^{-23}$ (J/K) is the Boltzmann constant. In our order of magnitude approach the volume $V \sim R^3$, where R is the characteristic dimension defining the volume. So our first expression for the gas pressure is, accordingly, $P \sim N T k/R^3$. Further, from Equation (3.2) we have that the central pressure varies to order of magnitude as $P_C \sim G M^2/R^4$, where $M = N m_p$ is the mass of the gas now assumed to be composed of hydrogen— as is appropriate for stars. In addition, we also note that if the average separation between the particles in our gas is d, then $N d^3 \sim$ volume $\sim R^3$. By equating the two expressions for the pressure we find a relationship for the central temperature such that $T_C = \alpha_G N^{2/3} (\hbar c/k d)$. Now, the first thing to notice about this expression is that it tells us that the central temperature must increase as the typical separation d between the gas particles decreases. The second point to notice is that once d is 'fixed,' the only other variable term is the number of particles in the gas N. All of the other terms are fundamental constants.

Let us now follow the gravitational collapse of a large, cold, pure hydrogen gas cloud. As the collapse proceeds, the volume of the cloud becomes smaller and the gas cloud heats up, since the spacing d between particles necessarily decreases. Eventually the temperature will become sufficiently high that the hydrogen will become ionized, resulting in equal numbers of protons and electrons being produced. Most of the mass of the cloud resides in the protons, since $m_p \sim 1000 m_e$. The electrons, however, can be squeezed together only so much during the collapse before Pauli's exclusion principle comes into play. Once the electrons become degenerate they can generate sufficient overcrowding pressure to halt the collapse of the gas cloud. So what we need now is to find an estimate of d_{min}, the minimum separation distance between electrons at which degeneracy becomes important. At this separation the temperature of the gas cloud will have reached a maximum value T_{max}.

Why Stars Are Massive

All material objects possess gravitational potential energy E_G, with $E_G \sim - G M^2/R$, where R and M are the radius and mass. As we have seen earlier, if no forces oppose the attractive gravitational force, then a body will collapse on the dynamical collapse timescale given in Equation (3.1). For our collapsing hydrogen cloud, however, the electrons will eventually become degenerate, and the overcrowding pressure resulting from the Pauli exclusion principle will halt the collapse. The point at which the contraction stops is determined by the condition that the gravitational potential energy per electron is comparable to the minimum kinetic energy of the electron K_0. This condition will allow us to find d_{min}. When $K_0 \sim (G M^2/R)/N$, we find by substitution that $d_{min} \sim 1/\alpha_G N^{2/3} (m_e c/\hbar)$.[29] So now we have an expression for the separation distance between electrons at the moment where degeneracy sets in. This result also allows us to determine the central temperature at the onset of degeneracy as $T_{max} \sim \alpha_G^2 N^{4/3} (m_e c^2/k)$.

We are now nearly at the point where an estimate of the minimum mass for a star can be established. Indeed, what has been found is that T_{max} is determined solely by the value of N, the total number of particles in the original gas cloud, all other terms in its evaluation being fundamental constants. As we saw earlier in this chapter, a compact, hot cloud of gas becomes a star (at least in name) once the central regions are hot enough for steady nuclear fusion reactions to begin, and this requires that $T_C \sim T_{NUC} \sim 10^7$ K. We also saw earlier in this chapter that once nuclear 'burning' begins the collapse of a star is halted, since it need not contract any more to replenish the energy radiated into space at its surface. In this respect the value of T_{max}, or more specifically N, determines the outcome of the collapse of a gas cloud. If $T_{max} < T_{NUC}$ then nuclear reactions will not be initiated in the cloud, and it will collapse until electron degeneracy sets in, at which point the cloud will simply begin to radiate its internal energy into space and cool off. If, on the other hand, $T_{max} > T_{NUC}$, then nuclear fusion reactions will have begun before degeneracy sets in, and the cloud will become a star. To a first approximation an estimate for the 'stardom' condition can be set as $T_{max} = T_{NUC}$, which requires that $N > 10^{56}$, which in turn indicates that the minimum mass for a star is $M_{min} = 10^{56} \, m_p \sim 0.1 \, M_\odot$. What this result tells us is that only assemblages of at least 10^{56} particles (i.e., hydrogen atoms) can possibly turn into stable stars with hydrogen fusion reactions occurring in their cores. Smaller assemblages with $N < 10^{56}$ particles will form stable degenerate bodies (i.e., brown dwarfs) held together by their own gravity, but supported by the electron overcrowding pressure resulting from the Pauli exclusion principle.

The Sun contains $N \sim 10^{57}$ particles, so it sits appropriately above M_{min}, but we now have to ask the next obvious question: "What is the greatest mass that a star can have?' We won't go through all the details here, but it turns out[26] that as T_{max} increases, so the contribution of radiation pressure P_{rad} becomes more and more important, and once the mass of a star is greater

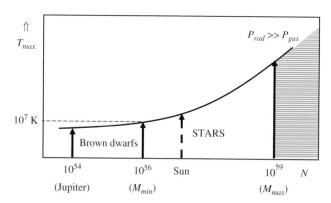

FIGURE 3.13. Schematic plot of the maximum temperature T_{max} obtained during collapse, against the number of particles N contained in the collapsing cloud. During collapse, each cloud moves vertically upwards in the diagram until it intersects the curve corresponding T_{max}. No stars can form in the shaded region to the right of the diagram due to the dominance of radiation pressure. Likewise, no stars can form for $N < 10^{56}$ since for these objects $T_{max} < T_{NUC} \sim 10^7$ k.

than $\sim 100 \ M_\odot$ (corresponding to $N > 10^{59}$ particles) an insta-
bility sets in with the result that a would-be star is disrupted.
Somewhat loosely, one can say that massive clouds 'bounce,' since
material is eventually driven outward, back into space, by the
strong radiation pressure. A summary of the possible star-forming
collapse scenarios is shown in Figure 3.13.

A Constraint on Planet Building

In addition to explaining the characteristic masses of stars, the
fundamental constants also set a constraint upon the existence of
planets. We won't derive the full result here because it is rather
complex, but what Professor Brandon Carter[30] has argued is that
if α_G were just slightly greater than its deduced value (based,
remember, on fundamental constants), then all stars would be
fully convective, low-temperature dwarfs. Further, if α_G were just
slightly smaller than its actual value, then all stars would be hot
with fully radiative interiors.

This is indeed a remarkable result. If all stars were fully
convective, low-temperature red dwarfs, then there would be
no supernovae and no production of carbon or other elements
essential to life. If all stars were hot and radiative, then it is
presently unclear if there would be any planets. The reason for
this latter claim is a little complex, but relates to the idea that
the spin rate of newly forming stars is linked to the development
of strong magnetic fields within their outer convective layers.
This magnetic braking effect is certainly observed. Hot, massive
stars with radiative envelopes spin rapidly, while cool, low-mass
stars (like the Sun) with convective outer layers spin slowly.
Not only this, current theories of planet formation require that
the material within the accretion disk around a newly forming
star sheds its angular momentum as it spirals inward. The most
likely way of doing this is via the formation of spin-axis aligned
jets constrained by magnetic fields. Indeed, such jets have been
observed in a number of newly forming star systems. In addition,
Luisa Rebull and co-workers at NASA's Spitzer Science Center[31]
recently conducted a study of some 900 stars in the Orion Nebula
and found that slow-spinning young stars are five times more likely

to have disks (in which planets might form) than fast-spinning young stars. There are many tenuous and currently unclear threads in the arguments just presented, but as Carter points out, "If this is correct, then a stronger gravitational [fine structure] constant would be incompatible with the formation of planets and, hence, presumably of observers."

Our discussion of the physics of stellar interiors is now complete. For those who have struggled through the mathematical arguments, well done! For those who want to see more details, then take another look at Reference 3. Hopefully, in the meantime, we have made clear some of the remarkable and elegant ideas underlying the modern-day theory of stellar structure and evolution. In the next chapter – before we move on to consider how a future star engineer might try to manipulate the properties of our Sun – we will look at the costs and consequences of not intervening in the Sun's aging process.

Notes and References

1. The spectral classification scheme is essentially a means of arranging and recognizing similar stars according to their surface temperature (as illustrated in Figure 3.8). The classification is based upon the characteristics of specific absorption lines recorded in stellar spectra. Importantly, the 'strength' of an absorption line varies with temperature. The classification scheme runs according to the designations O, B, A, F, G, K, and M (each designation having a set of subtypes—i.e., A1, A2…A5). The O and B stars are the hottest stars with atmospheric temperatures in excess of 15,000 K. The K and M stars have the coolest atmospheres, with temperatures varying between 5,000 and 3,000 K respectively. The Sun is a G spectral type star. More specifically, it is a G2 spectral type star, indicating that it has a temperature of about 6,000 K.
2. The highly recommended and well-produced Extrasolar Planet Encyclopedia can be found at: http://exoplanet.eu/.
3. A good introductory book on stellar structure and evolution is that by R. C. Smith, *Observational Astrophysics* (Cambridge University Press, Cambridge, 1995). C. J. Hansen and S. D. Kawaler provide a detailed technical description of the stars in their *Stellar Interiors:*

Physical Principles, Structure and Evolution (Springer-Verlag, New York, 1994). In a paper entitled Order-of-magnitude "theory" of stellar structure, [*American Journal of Physics*, **55** (9), 804–810, (1987)], George Greenstein develops, in a highly readable fashion, a series of analytic formulas for a model star. A number of the equations that are developed in this chapter and used in the next are explained in the paper: A novel stellar model: 'a sacrifice before the lesser shrine of plausibility' [M. Beech, *Astrophysics and Space Science*, **168**, 253–261, (1990)].

4. A. S. Eddington, *The Internal Constitution of the Stars*, Cambridge University Press, Cambridge (1926).

5. With these boundary conditions we are ignoring the fact that the Sun has a complex outer structure composed of the chromosphere and the corona. For the construction of most mathematical stellar models this is not a real problem, since the outer regions of a star contain relatively small amounts of mass.

6. The perfect gas equation can be used when the particles in a gas do not interact with one another. Although stars do have regions of very high density, the temperature of the stellar gas is so high that the atoms are broken down into their constituent nuclei and electrons. The typical spacing between components in the stellar gas is then much larger than the sizes of its individual components and, consequently, particle interactions are actually quite rare.

7. The atomic mass number refers to the total number of protons and neutrons in the atomic nucleus. In many atomic species there are equal numbers of protons and neutrons in the nucleus and, consequently, the number of electrons associated within a neutral atom will be half the atomic mass number.

8. The relative abundances of many of the Sun's constituent elements can be determined directly from the study and modeling of its spectrum. The actual abundances of many elements, however, are typically determined through the laboratory study of primitive (that is, unprocessed by heat) carbonaceous chondrite meteorites.

9. Detailed computer models indicate that the Sun is centrally condensed, with 50 percent of its mass being contained within a region enveloping just the central quarter of its radius. The outer convection zone, while extending over a third of the Sun's radius, contains about 5 percent of the Sun's actual mass.

10. For those who check the algebra there is actually a factor of three-quarters missing from Equation (3.9), but for this order of magnitude argument we have taken this to be sufficiently close to unity and can ignore the difference.

11. In addition to the PP chain, hydrogen can be converted into helium through the CNO cycle. In this case, four hydrogen nuclei are converted into a helium nucleus, two positrons, two neutrinos, and energy, with the help of a catalytic ^{12}C nucleus. Exactly the same amount of energy is generated by the CNO cycle as in the PP chain, but the CNO cycle only operates efficiently at temperatures somewhat higher than those found in the Sun. Stars more massive than about 1.3 $M_\odot$, in fact, have sufficiently high central temperatures (see Equation 3.5) that the energy generated by the CNO cycle is greater than that produced by the PP chain.

12. Even if deuterium was produced more rapidly, the PP chain would still be slowed down by the $^3He + ^3He$ reaction, which runs at a rate of about 10^5 years per reaction. The main sequence lifetime would nonetheless be significantly reduced (by some five orders of magnitude) and, as described in Chapter 2, this would prohibit the evolution of advanced life on Earth. We literally exist, in an Anthropic principle sense, because the first stage of the PP chain is so very slow.

13. The actual expression for the luminosity is $L \sim \mu^{7.5} M^{5.5} R^{-0.5}/(1 + X)$, but we have substituted for $R \sim$ constant x M. This correspondence actually follows from Equation (3.5), given that solar mass stars 'run' the hydrogen fusion reactions at a near constant central temperature.

14. The dramatic changeover in the mode of energy transport within stars relates partly to the dominant mode of energy generation. As stellar mass increases, so the CNO energy generation mode (see Note 11) becomes increasingly dominant, and its high temperature sensitivity results in the central core becoming convective. In this manner stars more massive than about 1.3 $M_\odot$ have convective cores and radiative envelopes, while stars with a mass of less than about 1.3 $M_\odot$ have radiative cores and convective envelopes.

15. Helium is 'converted' into carbon via the so-called triple alpha reaction: $3\ ^4He \Rightarrow\ ^{12}C + \gamma$. The helium atom nucleus, composed of two neutrons and two protons, is commonly called the alpha particle, and it is the interaction and fusion of three alpha particles that can produce a ^{12}C nucleus.

16. White dwarf stars are a stable (that is, very long-lived) end phase of stellar evolution. A solar mass white dwarf is of about the same physical size as Earth (or $\sim 1/100$ the size of the present-day Sun) and, consequently, it is an object with a very high internal density. Nobel Prize-winning astrophysicist Subrahmanyan Chandrasekhar showed, however, that there is a limit to how massive a white dwarf can be and, once the mass exceeds 1.4 $M_\odot$ it will catastrophically collapse as a Type II supernova, as discussed in Chapter 2.

17. If we define stars as being objects that can initiate hydrogen fusion reactions at some stage in their evolution, then a minimum mass of 0.08 $M_\odot$ is required to acquire star status. Brown dwarfs with masses, M_{BD}, such that 15 $M_{Jupiter} < M_{BD} < 0.08$ $M_\odot$ are neither stars nor large Jovian-like planets, but intermediate objects destined to eventually become low-mass black dwarfs. Brown dwarfs can undergo a deuterium fusion reaction phase $(D + P \Rightarrow {}^3He + \gamma)$, while Jupiter-like planets generate internal energy through gravitational contraction.

18. Neutron stars are objects that contain a solar mass of material in a sphere of radius 20 km or so. The exceptionally high density encountered within neutron stars results in the neutrons becoming degenerate and this quantum mechanical effect can support the star against gravitational collapse. There is, however, an upper limit to the mass of a stable neutron star and this is estimated to be between 2 and 3 solar masses.

19. Ironically, the greenhouse warming of Earth's atmosphere that is currently of great concern was of vital importance when the Sun was younger and less luminous. Without some additional atmospheric warming Earth's early oceans would have frozen over, substantially altering the global climate and presumably delaying the onset time for the first emergence of life.

20. See, for example, I-J. Sackmann, A. J. Boothroyd, and K. E. Kraemer, Our Sun III: present and future, *The Astrophysical Journal*, **418**, 457–468 (1993); P. Schrder, R. Smith, and K. Apps, Solar evolution and the distant future of Earth, *Astronomy and Geophysics*, **42**, 6.26–6.29 (2001).

21. One, of course, should never say infinite. Who knows what strange and currently unknown physics dictates the very long-term properties of matter. Nonetheless, it is currently thought that black dwarfs should remain stable for time periods many orders of magnitude greater than the current age of the universe.

22. In this argument we are assuming that there is a step-function, or jump, in the quantity ρ/μ at the core-envelope boundary. Detailed computer models, however, indicate that the composition varies over an extended zone. This, however, does not substantially alter the argument.

23. The reduction in the pressure scale height – the height over which the pressure falls by a factor of $e \approx 2.718$ – implies that the Sun's radius must increase for the pressure to vanish at its surface.

24. I have considered the effects of extreme central mass concentration in the article The formation of red-giants, *Astronomy and Astrophysics*, **156**, 391–392, (1986). One of the results from this study

was that the radius of a red-giant is determined (at least in part) by the mass of its central core. Peter Eggleton and R. C. Cannon [A conjecture regarding the evolution of dwarf stars into red-giants, *Astrophysical Journal*, **383**, 757–760 (1991)] have further argued that stars swell up in response to the development of a composition gradient produced by a hydrogen-shell burning source situated at the outer boundary of a star's inert core. A. P. Whitworth has 'experimented' with detained numerical models and in his article, Why red-giants are giant [*Monthly Notices of the Royal Astronomical Society*, **236**, 505–544, (1989)] finds that in addition to a molecular weight jump, increased central mass concentration and the presence of a hydrogen-burning shell, the opacity variation in a star's envelope is highly important in the production of an extended radius. See also the more recent article by Daiichiro Sugimoto and Masayuki Fujimoto, Why Stars Become Red-Giants, *Astrophysical Journal*, **538**, 837–853, (2000).

25. All of the carbon and oxygen atoms that exist in our universe were made inside of massive stars. The carbon atoms are produced through the triple-alpha reaction $3\ {}^{4}\text{He} \Rightarrow {}^{12}\text{C}$ + energy (see Note 15). The only reason this reaction actually 'works,' however, is because of the presence of an excited state of ${}^{12}\text{C}$ at the end point of the triple-alpha reaction. That this coincidence exists is entirely remarkable and, as Mario Livio and co-workers comment in their paper, The anthropic significance of the existence of the excited state of ${}^{12}\text{C}$ [*Nature*, **340**, 281–284 (1989)], "[As] is consistent with the anthropic principle, the energy of the resonant level of ${}^{12}\text{C}$ is required to have the value it does, to ensure carbon production and the consequent development of carbon-based life."

26. One of the very best shorter reviews on this topic is by B. J. Carr and M. J. Rees, The anthropic principle and the structure of the physical world. *Nature*, **278**, 605–612 (1979). See also the less mathematical article by John Gribbin and Martin Rees, Cosmic coincidence, *New Scientist* magazine, 13 January, 51–54 (1990)

27. Technically, the Pauli exclusion principle applies to so-called spin ½ particles, which includes electrons and neutrons, but not protons. Pauli's exclusion principle is the reason why atoms have different properties. Without the action of the exclusion principle all electrons would reside in the lowest energy ground state and chemistry as we know it would not exist.

28. There is a limit to the support that degenerate electrons can provide. If the characteristic speed of the electrons becomes relativistic (i.e., $V \sim c$), then their energy rather than varying as $p^2/2m_e$ will vary as p

c, with the result that a limiting mass M_C is imposed. If $M > M_C$ then collapse becomes inevitable. It turns out, as shown by Subrahmanyan Chandrasekhar (see Note 16), that $M_C \sim m_p/\alpha_G{}^{3/2} \sim 1\,M_\odot$.

29. From the condition $K_0 \sim (G\,M^2/R)/N$ we have $\hbar^2/(2\,m_e\,d_{min}{}^2) \sim G\,N^{2/3}\,m_p^2/d_{min}$ which then rearranges to the expression given in the text. Note, however, that the 2 has been dropped from the expression since it is a small number.

30. Brandon Carter, Large number coincidences and the anthropic principle in cosmology, in *Confrontaion of Cosmological Theories With Observational Data*. M. S. Longair (Ed.). IAU Symposium No. 63. Reidel, Holland (1974). pp. 291–298. Carter shows that normal stars fall between the two extremes of being either cool and fully convective, or hot and radiative throughout their interiors, provided $\alpha^{12}\,(m_e/m_p)^5 \sim \alpha_G$, where α is the electromagnetic fine scale constant. This condition, in our universe, is only just satisfied with the two sides having numerical values of 2.3 x 10^{-39} and 5 x 10^{-39}, respectively.

31. L. Rebull et al., A correlation between pre-main sequence stellar rotation rates and IRAS excesses in Orion, *Astrophysical Journal* **646**, 297–303 (2006).

4. The Price of Doing Nothing

Cost (noun). The price to be paid.
Cost (verb intransitive). Result in the loss of.
(from The *Concise Oxford English Dictionary*)

If our descendants do absolutely nothing about the aging of the Sun, then the future is clear: all life on and inside Earth will die. Indeed, all that will remain after the Sun has become a red-giant will be a sterile and heat-blasted Earth. Venus will possibly survive against destruction in the Sun's red-giant envelope, but Mercury is definitely doomed, and it will be consumed. The fate of our future desolate Earth will be to orbit in endless silence around a slowly fading white dwarf star.

In this chapter we will describe some of the costs of allowing the Sun to become a luminous and bloated red-giant. It is one possible future for our Solar System that we shall explore over the following pages, but it is not an inevitable future. Our descendants do, in fact, have a choice concerning their destiny.

The Habitability Zone

Earth is located in a very specific 'sweet zone' within our Solar System. Indeed, Earth is heated by solar radiation to just the right level that liquid water can exist on its surface. It is neither too hot nor too cold over most of Earth's surface. If Earth had formed much closer in toward the Sun, then all the oceans would have rapidly boiled away; if Earth had formed much further out from the Sun, then all of the oceans would have frozen. Venus and Mars are the possible alternatives to present-day Earth; one too hot for oceans, the other (now) too cold.[1] The region within the Solar System where liquid water can exist on the surface of a planet with an atmosphere such as that surrounding Earth is known as the *habitability zone*. At a distance d astronomical units from the Sun the surface temperature of a planet $T_{surface}$ expressed in Kelvins will be[2]

$$T_{surface} = T_{GHE} + 278 \left(\frac{(1-A)\,L(t)}{\varepsilon d^2 (AU)} \right)^{1/4} \tag{4.1}$$

where T_{GHE} is the greenhouse heating effect due to the atmosphere, A is the atmospheric albedo, ε is the emissivity of the planet,[3] and $L(t)$ is the luminosity of the Sun at time t expressed in terms of the Sun's present luminosity[4] [that is, $L(t = 4.5$ billion years) = 1]. The inner boundary of the habitability zone is determined by the boiling point of water, while the outer boundary corresponds to the distance at which water will freeze. Accordingly, at the present time within our Solar System the inner and outer boundaries for which $373 > T_{surface} > 273$ are $0.6 < d$ (AU) < 1.1 [derived from Equation (4.1) with $A = 0.2$ and $\varepsilon = 0.8$]. More detailed calculations, including model atmospheres that take active weathering reactions (which regulate the atmospheric CO_2 concentration) into account,[5] find that the habitability zone for our Solar System is actually shifted outward, falling somewhere between $0.95 < d < 1.4$. This latter range makes sense in that the habitability zone straddles Earth's orbit, but excludes both Venus and Mars as being too hot and too cold, respectively, for liquid water to exist at the present time.

The idea of a habitability zone can be applied to any parent star,[6] but as seen in Chapter 3, the luminosity of a star varies according to its mass [see Equation (3.12)]. The boundaries of the habitability zone will, therefore, be shifted inward for stars less massive than the Sun, and shifted outward for stars more massive than the Sun. Although no Earth-like extraterrestrial planet has so far been detected around another Sun-like star, a study by Brian Jones and co-workers[7] has investigated the orbital stability of hypothetical Earth-like planets within the habitability zones of stars known to have accompanying Jupiter-type planets. Among the systems that were studied it was found that the stars ρ CrB and 47 UMa could, in principle, have Earth-like planets in stable orbits situated in the habitability zone. Such planets could, again in principle, be capable of supporting life. Darren

Williams and co-workers[8] have also looked at the possibility of hypothetical moons in orbit around known extrasolar, Jupiter-like planets falling within a system's habitability zone. They find that the planets in the 47 UMa and 16 Cyg B systems could possibly support moons with liquid water at their surfaces (provided, that is, the moons are large enough to maintain an atmosphere).

The extrasolar planetary system HD 69830 is of particular interest with respect to both the deduced structure of the planets and the location of the planets around the parent star. HD 69830 is about 12.5 pc away, has a mass of 0.86 $M_\odot$, and hosts three Neptune-mass planets. Extensive computer modeling[9] has led to the suggestion that the innermost of the planets – with a mass about 15 times that of Earth and an orbital radius of 0.08 AU – has a rocky composition. In many ways, it is a super-sized Earth. The outermost planet, which is probably more similar to our Neptune in structure (i.e., a rocky/ice core surrounded by an extensive gas envelope), has an orbital radius of 0.63 AU, and this places it close to the inner edge of the habitability zone for the system. Recent observations with the Spitzer Infrared Telescope also indicate that the system sports what appears to be an asteroid belt in the region between 0.3 to 0.5 AU from HD 69830. This is actually an infrared emission excess due to small dust grains that the Spitzer telescope observations record, but since the system is estimated to be at least several billion years old, the dust is most probably a product of numerous asteroid collisions. With three planets moving in circular orbits – one of which is located near the habitability zone and an asteroid belt – HD 69830 shares many properties with our Solar System, and life on a large moon in orbit around the outermost planet might be possible (see Figure 4.1).

The Ocean on Europa

Europa is the second innermost of the four large Galilean moons that orbit Jupiter. It is about the same size as Earth's Moon, but rather than being a purely rocky body, Europa has an outer ice surface that caps a global ocean. The exact internal structure of Europa is not known, but various models indicate that it has an iron/iron-sulphide-rich core occupying about 30 percent of its interior.[10] Most of the remaining interior is taken up by a silicate mantle, but the outermost few hundred kilometers appear to be composed of a global, possibly salty ocean covered by a solid-ice veneer perhaps a few kilometers in thickness.

Equation (4.1) indicates that at 5.2 AU from the Sun the surface temperature of Europa is something like 120 K, well below the freezing point of water or brine (which freezes at a lower temperature because of its salt content). And yet, Europa has a global ocean! How is this possible? First, there is little doubt that there is a global ocean, since the *Galileo* spacecraft clearly detected a peculiar magnetic anomaly when it flew past Europa in December 1996. The most reasonable explanation of the recorded anomaly is that Europa has a near-surface conducting layer, and this is where the brine comes in. Next, of course, the question to answer is why hasn't the ocean frozen, since Europa formed along with the rest of the Solar System some 4.56 billion years ago. The reason why the ocean still exists is, in fact, remarkable, and it reminds us that liquid water can be found in locations well outside of the habitability zone defined – admittedly conservatively – above. The ocean of Europa hasn't frozen because of a tidal heating effect related to the non-circular orbit of Europa and the corresponding periodic stretching and relaxing that it undergoes in the strong gravitational field of Jupiter. This flexing and relaxing actually heats the outer part of its rocky mantle, and it is this heat that keeps the ocean from freezing.

Although there is little doubt that there is some form of global ocean under Europa's outer ice cap, it is far from clear if it can support life. It is likely that all the basic chemical ingredients to support primitive life are present in Europa, but by far the greater problem is how any life forms might produce the energy needed for their survival. Certainly the outer ice cap precludes photosynthesis from operating, so other chemosynthesis forms of energy generation will presumably need to apply. One possibility is that Europa might support isolated colonies of animals similar to those found around hydrothermal vents in Earth's deep oceans.

We do not currently know if life has gained a toehold in Europa's ocean, but spacecraft missions may provide us with an answer within the next three or four decades. In the meantime there is a place on Earth – Lake Vostok in Antarctica – which might provide us with a few clues as to what future Europa landers might find. Lake Vostok is several hundred meters deep, but more importantly, it is buried under some 4-km of ice. The icecap is estimated to be at least 500,000 years old, and it has been suggested that the lake may have preserved life forms that are significantly different from those found anywhere else on Earth. An active core-drilling project has been conducted at the Lake Vostok site, and the bore-hole is estimated to be very close (a few tens of meters) to breaking through into the liquid layer. The final penetration effort, however, is currently on hold, as engineers try to ensure that no, or at least a minimum, of pollutants (i.e., the drilling fluid) are introduced into the pristine waters of the lake.

A Brief Aside: Utilizing Europa

If no life forms are detected within Europa's ocean, then one good use for this vast brine resource would be to 'seed' it with appropriately engineered (genetically engineered, in this case) halophiles—equivalents of the salt-loving extremophiles found, for example, in the Dead Sea in Israel. If such seeding and successful breeding can be nurtured, then a potentially massive food resource could be cultivated.

Moon Life

The current models accounting for the formation of moons orbiting large gas-giant planets, such as Jupiter, suggest that they form in naturally occurring circumplanetary disks. There is no specific reason, therefore, why the Jupiter-mass planets being discovered around other Sun-like stars shouldn't also have moons. The question concerning how far moon-based life might evolve is completely open at the present time. In the case of Europa one might reasonably imagine that microbial life could have evolved, but it is not at all clear if intelligent life, capable of, say, manipulating its environment or even colonizing an entire planetary system is possible.

Synchronization and the Moon Effect

Figure 3.9, in the last chapter, indicates that stars can form with masses as small as a few tenths of a solar mass, and it is also known that these stars can support planetary systems. The low temperature, low luminosity, M dwarf star Gliese 581 is one especially intriguing example of a low-mass [0.3 $M_\odot$] star that has three known planetary companions. The system holds great interest since one of the planets is a so-called 'super-Earth' planet. The planet is about 1.5 times larger than the Earth (five times more massive) and orbits Gliese 581 in just under 13 days. Although located very close to Gliese 581 (at a distance of 0.07 AU) the planet is nonetheless in the system's habitability zone, and Stephane Udry, of the Geneva Observatory, and co-workers have recently suggested that the planet might have regions where surface water could exist.

Even though the system is estimated to be some 4.3 billion years old, it is unlikely that the 'super-Earth' companion to Gliese 581 supports any life (or, for that matter, any extensive oceans). The reason for this latter statement is exemplified by our Moon, which is in synchronous rotation around Earth. Due to the close proximity of the Moon and Earth, and the fact that the Moon and Earth are not perfect spheres, the Moon's spin rate has been brought into equalization with its orbital motion. This is why we always see the same face of the Moon from Earth. Since low-mass stars are also low luminosity stars, their habitability zones are located close in toward the star. Indeed, the habitability zones will be so close to the parent stars that planetary synchronization will inevitably come about.

For our Moon synchronization is not a problem, but for a planet with an atmosphere the effect will most certainly be catastrophic. Since one hemisphere of the planet will always face the parent star it will be continuously warmed, while the other hemisphere will be constantly cooled by radiating its energy into outer space. The outcome of such extreme heating and cooling is not absolutely clear, but most meteorologists suggest that the atmosphere will eventually freeze out. This being said, some researchers have argued that stable atmospheres capable of supporting regions of surface water might still form around synchronized planets. Detailed numerical modeling indicates that planets located within the habitability zones of stars less massive than 0.5 $M_\odot$ will be synchronized, and we have therefore taken this to be the lower stellar mass limit for supporting a habitable planet. The main-sequence lifetime of a 0.5 $M_\odot$ star is about eight times greater than that of the Sun's, so habitable planets in orbit around these stars have a naturally extended lifetime.

The Upper Limit

Stars are observed to form with masses perhaps as high as 100 times that of the Sun. Such stars are very rare, but they do form. Where, then, might we place the upper limit to the mass of a star capable of supporting planets on which intelligent life might evolve?

As discussed in Chapter 1 the mass limit can be set (to a first approximation) according to the main-sequence lifetime of the star being longer than the time required for human beings to have appeared on Earth: T_{US} = 4.5 billion years. According to Equation (2.1), this sets a limit of about 1.3 $M_\odot$ on the mass of a star capable of supporting intelligent life. Life may well evolve on planets within the habitability zones of stars outside of the chosen mass range of 0. 5 to 1.3 $M_\odot$, but at this stage it can be assumed that such life is most likely microbial in nature rather than of an advanced form capable of space exploration.

Although massive stars have relatively short main-sequence lifetimes this does not preclude the possibility of them having planets. The very high luminosities associated with such stars dictate that the planets cannot be too close in – or else they would literally boil away – but there is some limited, albeit controversial, evidence that they do at least form. Not only is there evidence that planets form around massive stars, there is also – again controversial – evidence that massive stars can consume their planetary progeny. This latter possibility is based upon recent interpretations of the 'outburst' observed from the star V838 Monocerotis (Figure 4.2). V838 Mon is a binary system composed of two ~8 $M_\odot$ stars situated between 6 to 10 kpc away from the Sun. In early 2002 the system underwent a series of distinct outbursts in brightness over a period of about 100 days. Three outbursts were actually recorded, with each event lasting about 25 days. As a consequence of this sudden increase in brightness a pulse of light spread outward from V838 Mon into space, illuminating in its path the surrounding interstellar medium as seen in the dramatic Hubble Space Telescope image reproduced in Figure 4.2 A number of distinctly different explanations for the origin of the outbursts have been published, but Alon Retter (Penn State University) and co-workers[11] have argued that the three outbursts were the result of one of the stars in V838 Mon consuming three Jupiter-sized planets. Detailed numerical modeling suggests that as a consequence of devouring the first planet, the host star expanded and this resulted in the envelope entrapment and eventual consumption of the next two planets. Provided this model for the outburst of V838 Mon is the correct one (the debate still rages), it tells us that not only can planets form around massive stars, but that planets can also form in binary star systems.

A Moving Habitability Zone

As the Sun's luminosity steadily increases with time, Equation (4.1) tells us that the surface temperature of the different planets must also increase, thereby shifting the habitability zone outward and deeper into the Solar System. Using the solar evolution model described in Chapter 3 [see Table 3.2], the location of the hot, inner boundary for the habitability zone is shown in Figure 4.3. Here we have used Equation (4.1), which while oversimplified for the problem to be discussed is nonetheless representative of the point. And the point is, as the Sun ages, so the inner (hot) boundary of the habitability zone is swept into the outer reaches of the Solar System and, indeed, once the Sun enters its asymptotic giant branch phase, even the icy moons of Saturn will begin to evaporate.

The simplified model calculations appear to indicate that Earth's oceans will begin to boil away in about 7 billion years. In fact, the situation is more urgent than that and the oceans will, in fact, enter a significant evaporation phase in about 1 billion years. The reason for the accelerated demise of the oceans is due to the evolution of Earth's atmosphere,[12] which is not taken into account

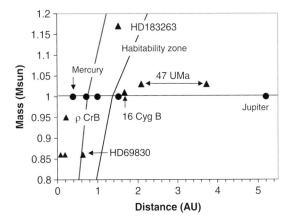

FIGURE 4.1. The habitability zone for stars in the range 0.8 to 1.2 times the mass of the Sun. The inner planets within our Solar System are shown (circles), the location of the planets within a selection of extrasolar systems (triangles) are also indicated. Based upon the calculations by Kasting and co-workers.[5]

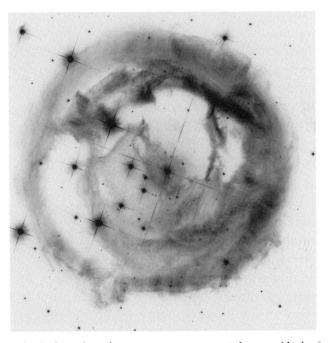

FIGURE 4.2. The light echo of V838 Monocerotis. A burst of light from V838 Mon has illuminated the dust in the surrounding interstellar medium. (Image courtesy of the Hubble Space Telescope Institute and NASA)

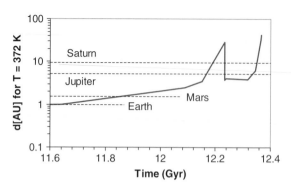

FIGURE 4.3. Evolution of the inner hot edge of the habitability zone within our Solar System. Note that the boundary line [calculated according to Equation (4.1)] is the actual surface temperature of the planet, assuming that it has no atmosphere and ocean. Earth's dry surface temperature will reach 100 °C (372 K) in about 7 billion years; the surface of Mars will reach this temperature about 200 million years later. The moons of Jupiter and Saturn will first fall in the 100 °C heating zone in about 8 billion years from now as the Sun ascends the giant branch.

in the calculations leading to Figure 4.3. Likewise, life will have been killed off long before Earth's equilibrium temperature reaches 100°C. Indeed, only the most extreme of bacterial life forms – the so-called extremophiles[13−] can survive and potentially thrive in environments where the temperature exceeds 40°C for prolonged periods of time. There is a clear symmetry to the story of life on Earth when we compare the deep future to the distant past: microorganisms were the very first life forms to appear on Earth, and they will be the very last to die.

The Beginning of the End

Although it is the Sun's increasing energy output (that is, its luminosity) that will eventually cause Earth's oceans to evaporate and kill off even the hardiest of life forms, it will be the Sun's increasing size that will destroy Mercury. In about 8 billion years from now, when the Sun begins to ascend the red-giant branch (points 3 to 4 in Figure 3.10), it will swell up to engulf the entire orbit of Mercury, which orbits the Sun at a distance of 83 solar radii.

Inevitably, as the Sun's radius expands outward, Mercury will find itself moving through an increasingly dense gas. The planet will then begin to accrete material from the Sun's envelope, and it will begin to experience the drag effects associated with its motion through the Sun's extended atmosphere. As we shall see below, Mercury will be rapidly destroyed as it falls deeper and deeper into the Sun's envelope. For a short few years it will circle the Sun like a glowing meteor, with its vast bulk eventually being broken apart and ablated into its constituent atoms.

If the accretion rate of material by Mercury from the Sun's envelope is written as M_{acc}, then the characteristic time for its orbit to decay will be $T_{decay} \approx M_{Merc}/M_{acc}$, where M_{Merc} is the mass of Mercury.[14] The accretion rate can be estimated as being of the order of the cross-sectional area of Mercury $(\sigma = \pi R_M^2)$ multiplied by the density of the Sun's envelope at the orbit of Mercury (ρ_{env}) multiplied by Mercury's orbital velocity V_{Merc}. Mercury's orbital speed is V_{Merc} = 48 km/s, its radius is R_M = 2,440 km, and its mass is M_{Merc} = 3.3 x 10^{23} kg. Taking an envelope density[15] of ρ_{env} = 10^{-4} kg/m^3, an orbital decay time of $T_{decay} \approx$ 100 years is indicated for Mercury. The orbital decay time will probably be more rapid than this estimate because we haven't included the effects of gravitation, which will tend to increase the accretion rate.[16] Therefore, once the Sun starts to climb the red-giant branch, Mercury will begin to spiral inward on its orbit, speeding ever more rapidly toward a fiery destruction.

The Fate of Venus and Earth

Although Mercury is doomed once the Sun becomes a red-giant, it appears that Venus and Earth, at least as physical entities, will survive. The ultimate fate of these planets, however, will reside in the amount of mass lost by the Sun during both its red-giant and asymptotic giant branch phases. The more mass our future Sun ejects into space the more likely it is that Venus and Earth will be spared from a fiery obliteration. The reason for this dramatic escape is due to the fact that as the Sun loses mass, so the orbital radius of each of the planets will increase, carrying them

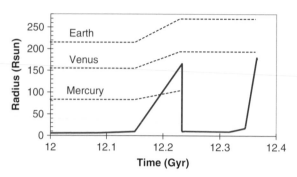

FIGURE 4.4. Changes in the Sun's radius with age. Mercury will be consumed by the Sun in about 8 billion years from now. Note that in this particular set of calculations both Venus and Earth survive against destruction because the Sun loses mass (about 0.2 $M_\odot$) as it ascends the giant branch. [Solar model based upon calculations by Sackmann, Boothroyd, and Kraemer, *ApJ.* **418**, 457–468 (1993)]

beyond the reaches of the Sun's bloated envelope (as shown in Figure 4.4).

At the present time Earth's orbital radius (1 AU) is equivalent to 215 $R_\odot$. Detailed computer models predict that the Sun will expand to somewhere between 200 to 250 times its present size during its red-giant phase which, without orbital migration, suggests that Earth will be destroyed. The change in Earth's orbital radius as a result of the Sun losing mass can be determined by assuming that Earth conserves its orbital angular momentum.[17] In this fashion, Earth's orbital radius $a(t)$ can be related to the Sun's mass $M_\odot(t)$ at some future time t, as $a(t) = a(0)$ $M_\odot(0)/M_\odot(t)$, where $a(0)$ and $M_\odot(0)$ are the values of Earth's orbital radius and the Sun's mass at some specified starting time $t = 0$ (i.e. now).

For Earth to survive against the extreme expansion predicted for the Sun's outer red-giant envelope we require $a(t)/a(0) \approx 250$ $R_\odot/215\ R_\odot = 1.16$ at the time that the Sun reaches the red-giant tip (Point 4 in Figure 3.10). In other words, the Sun's mass must decrease by a factor of $M_\odot(t)/M_\odot(0) = 0.86$ for Earth to be sure of escaping consumption. For Venus to survive, $a(t)/a(0) \approx 250$ $R_\odot/155\ R_\odot = 1.61$, and correspondingly, $M_\odot(t)/M_\odot(0) = 0.62$. Observations clearly indicate that solar-mass stars do lose about 0.2 to 0.3 $M_\odot$ during their red-giant branch phases. Earth, it therefore appears, is probably safe from physical destruction during the red-giant branch and asymptotic giant branch phases of the Sun's evolution. The detailed computational models of solar evolution, however, are not currently in agreement as to whether the Sun will lose enough mass for Venus to survive. Calculations by Peter Schrder[18] and co-workers at the University of Sussex, for example, predict that the future mass-loss of the Sun will result in the survival of Earth, but not of Venus.

The Outer Planets

Moving into the outer Solar System, beyond the orbit of Mars, the Sun's increasing luminosity is not likely to have any great effect upon the internal structure of the Jovian planets. Both Jupiter and Saturn have massive hydrogen and helium envelopes, and these can adjust without any great structural changes to accommodate the increased energy output received from the aging Sun. Examples of the possible future atmospheric states of Jupiter and Saturn can be found among the extrasolar planets. Indeed, many of the extrasolar planets have exceptionally small orbits, and this means that they are heated to temperatures in excess of those experienced by the planets in our Solar System. The most extreme case known is that for the 1.45 Jupiter mass planet OGLE-TR-56b, which has an orbital radius of 0.0225 AU. This planet orbits its parent star in an incredible 1.2 days at a distance 17 times closer than Mercury orbits our Sun, resulting in a surface temperature of about 1,500 K. In the few cases where measurements have been made, the hot Jupiter-like planets appear to have diameters that barely differ from those of their cooler counterparts orbiting at much greater distances from their parent stars. It is upon this basis that we would not expect Jupiter or Saturn to experience any extensive internal changes as the Sun ages. This being said, it is highly likely that the appearance of their upper cloud decks will change dramatically. The enhanced solar heating will warm the atmosphere, driving stronger zonal winds and altering the details of the photochemical reactions responsible for producing the various colored spots, ovals, and bands.

In similar fashion to their gas-giant cousins, the icy giant planets Uranus and Neptune will ride out the effects of the Sun's increasing luminosity essentially unscathed. Again, the main observational changes will be cosmetic and relate to the extra heating of their upper cloud decks and outer atmosphere. For a short few hundred million years – some 7.5 billion years from now – Uranus will be located in the Solar System's habitability zone (see Figure 4.3). During this time an interesting possibility for terraforming arises. More correctly perhaps, one should say aquaforming in the case of Uranus, since the idea here is to generate a water world—literally, a planet encircled by a deep global ocean. Alain Léger of the Institu d'Astrophysique Spatiale in Orsay, France, and co-workers have suggested that among the extrasolar planetary systems, a family of 'ocean-planets' might well exist.[19] These worlds will be smaller in mass than Uranus, by a factor of about two, but will have a similar internal structure. Essentially, and in contrast to Uranus, what they lack is an extensive hydrogen atmosphere. Léger and co-workers have found that ice-rich planets with masses of between 6 to 8 times that of Earth, situated in the habitability zone of their parent star, can develop global oceans up to about 100 km in depth.

Models for the interior structure of Uranus are not especially well constrained at the present time, but it does appear that the planet has a central rocky core and an extensive icy mantle. Indeed, the icy mantle accounts for about 80 percent of the mass of Uranus (a total of 11.5 Earth masses of material), and the core accounts for a further 7 percent (about 1 Earth mass of material). To trim Uranus down to an aquaforming mass, something like 6 Earth masses of material will have to be removed from the planet's atmosphere. This removal might be in the form of direct mining, since the extracted hydrogen, helium, oxygen, and nitrogen could be used in numerous industrial processes (see Chapter 6). The mass removal process could also be more dramatic (and wasteful) with the excavation being driven by multiple impacts from Kuiper Belt objects specifically maneuvered into position. An appropriately aquaformed Uranus could support structures such as the supramundane platforms described by Paul Birch (see Note 40 in Chapter 2), and it could provide a near inexhaustible supply of fluids for an intensive hydroponics industry.

Orbital Engineering

Irrespective of what the numerical calculations for the Sun's future evolution predict, our future descendants will likely feel that the planet Venus is worth saving, especially if it has already been successfully terraformed. Likewise, our descendants may feel that Earth itself requires an additional safety zone, such that it is placed well beyond the outer reaches of the (non-engineered) red-giant Sun's envelope. There are, in fact, ways in which this outward orbital migration might be achieved. In addition to reacting to any mass lost by the Sun, planetary orbits can be adjusted via repeated close gravitational encounters with smaller objects. Don Korycansky and co-workers[20] have outlined, for example, a scenario by which orbital energy can be transferred from Jupiter to Earth, thereby increasing Earth's orbital radius. And, as suggested above, a similar process of orbital engineering might be used to aquaform Uranus.

In the Korycansky, et al scheme, a ~100-km diameter asteroid or Kuiper Belt body is used as an energy transfer object. The idea is that the smaller body first gains energy via a gravitational slingshot encounter with Jupiter. That gained energy is then deposited into Earth's orbital motion by an appropriately controlled close, leading limb flyby. By repeating this process every 6,000 years or so for the next billion years the Korycansky team believes that Earth can be maneuvered into an increasingly large orbit such that its equilibrium temperature [as given by Equation (4.1)] remains constant throughout the Sun's main-sequence lifetime. Accordingly,[21] in a billion years from now, Earth's orbital radius will need to have increased to 1.045 AU; 4 billion years from now Earth's orbital radius will need to be 1.225 AU.

Korycansky and co-workers point out that "Any serious proposal for planetary engineering, or any large-scale alteration of the Solar System, raises important questions of responsibility."[20] Absolutely! They also point out that their proposal has a number of potential problem points. The transfer of energy from one object to another via gravitational assists is well understood, and it is already a routine part of spacecraft navigation. The problem, however, is if Earth is gaining orbital energy from Jupiter, and thereby moving slowly outward, Jupiter is losing orbital energy and slowly moving inward. Because of Jupiter's size, it doesn't actually move very far – about 0.01 AU closer to the Sun over the Sun's main-sequence lifetime. This inward shift is small, but Jupiter is very close to the outer edge of the main-belt asteroid region and, consequently, the inward drift might destabilize the orbits of numerous asteroids. It is possible, therefore, that the orbital migration program might enhance the NEA population of asteroids and consequently increase the asteroid impact problem. In addition, it is not clear what happens to Earth's Moon in the Korycansky scenario. Since the Moon's gravitational influence is vital for stabilizing the obliquity of Earth's spin-axis, its loss could result in dramatic and chaotic swings in the climate.[22] The fate of the planets Venus and Mars are not presently resolved in the scenario outlined by Korycansky and co-workers, but presumably the basic process could be multiplied to adjust their orbits as well. However, one starts to feel a little uncomfortable at the sheer complexity of trying to simultaneously manipulate the orbital radii of three of the terrestrial planets in such a fashion that Solar System stability is maintained and unwanted collisions do not occur. The Korycansky scheme, as currently outlined, would require about a million close Earth flybys by a ~100-km diameter object; one slip, and Earth is sterilized as effectively as not increasing Earth's orbit at all. With this possibility of catastrophic disaster in mind, Colin McIness has suggested[23] that a large reflective sail (Figure 4.5), suitably stabilized near Earth, could be used to increase the size of Earth's orbit through the action of solar radiation pressure. This method avoids having to coordinate the numerous close flybys that the gravity-assist scenario calls for and alleviates the chances of a catastrophic impact by accident. The McInnes scenario requires the construction of a 5×10^{16} m^2, 8 μm thick, 10^{15}-kg mass metallic solar sail from an 9-km diameter M type (i.e., a nickel-iron rich) asteroid. Such a sail (which corresponds to a disk of radius 19.2 Earth radii), if maintained at a stand-off distance of 300 Earth radii, could enlarge Earth's orbit to 1.5 AU over a time interval of about 6 billion years. Leonid Shkadov[24] has outlined an even grander scale use for very large solar sails and suggests that the entire orbit of the Solar System about the galactic center could be controlled. Another use for such reflectors – also called class A stellar engines – is, of course, to deflect nearby stars

FIGURE 4.5. An artist's rendering of the *Cosmos-1* space sail. Developed for the Planetary Society, an unfortunate launch-rocket malfunction resulted in the space sail never reaching an operational orbit. In the future, however, solar sails will probably be employed to manipulate the orbit of asteroids, comets, and Kuiper Belt objects. Image by Rick Sternbach and the Planetary Society.

from passing too close to our Solar System's Oort Cloud, thereby triggering a potentially deadly cometary shower (see Chapter 2).

The orbital manipulation of diverse objects within the Solar System will presumably play an important part in our distant descendants' strategy toolkit for long-term survival. The gravitational assist method, for example, could be exploited to not only avoid Earth's heat death, but to aid in the terraforming of Mars by (initially) nudging its orbit closer in toward the Sun. Likewise, the process could be used to clear potentially impacting asteroids from near-Earth space, or maneuver asteroids, Jupiter-family comets, and Kuiper Belt objects into orbits where they could be more easily (and safely) mined for their resources. The orbital shepherding of these same objects might also produce 'useful' planetary impacts. The bulk of Venus's overburdened atmosphere, for example, could be removed by repeated large body impact, thereby allowing the terraforming process to begin.

What is perhaps most remarkable about the scenarios outlined above is that the technology and know-how to complete the tasks described already exists. The material to make working space sails has been developed, and small-scale space sails and inflatable structures have also been successfully deployed in near-Earth orbit. We are literally on the cusp of taking space sail technology into the Solar System now, and the first steps toward the more challenging engineering projects that our descendants will want to make are already being taken.

Waving the Flag

In addition to the shepherding and orbital manipulation of asteroids, cometary nuclei, Kuiper Belt objects, planets, and perhaps the entire Solar System, our descendants may even build massive solar sails just to advertise humanity's existence. The same idea, of course, might also occur to extraterrestrial civilizations. Indeed, Luc Arnold of the Observatoire de Haute-Provence in France has suggested[25] that SETI searches could be established on the basis of looking for unusual transit features. If, for example, an extragalactic civilization constructs a large solar sail having a triangular shape or a louvered structure, then the light curve produced by the sail each time it passes in front of the system's parent star would be readily detected by a diligent observer (provided, of course, that the transit geometry was favorable for the observer).

With respect to the required resources to build such structures, Arnold comments that a 12,000-km diameter, 1-micrometer thick solar sail made of iron would have a mass of about 10^{12} kg. While this certainly seems large, it is equivalent to the material contained within a 632-m diameter iron asteroid. Equivalently, as noted by Arnold, 10^{12} kg roughly corresponds to the annual production of iron on Earth. Once again, what is remarkable about the construction of such massive structures is that the know-how and the resources to produce them already exist. All that humanity lacks at the present time is the inclination to start the project.

End Games and Exotic Worlds

It is not just the inner Solar System that will be disrupted if our Sun is allowed to become a red-giant. The outer Solar System will also feel the effects of the Sun's growing luminosity. Although, as we said earlier, the Jovian planets won't be affected much by the Sun's increasing luminosity, their many attendant moons will be, since water ice is a major part of their internal makeup. As indicated by Equation (4.1), as the Sun's luminosity increases, so the heliocentric distance at which water ice begins to rapidly sublimate moves deeper and deeper into the Solar System. Literally, the boundary of an expanding sublimation sphere will sweep through the outer Solar System as the Sun ages to become a red-giant. Inside of this sphere ice will begin to evaporate rapidly. Currently the ice evaporation boundary is situated some 1.5 to 2 AU from the Sun. It is upon reaching this boundary, for example, that we see tails and extended comas begin to appear around cometary nuclei—this coming about because cometary nuclei are predominantly composed of water ice. As the Sun's luminosity increases, however, the ice evaporation boundary will move deeper into the Solar System. Eventually the Galilean moons of Jupiter will begin to lose their surface ices, the atmosphere of Saturn's Titan will boil away, and the innermost Kuiper Belt objects, Pluto, and Charon, will begin to develop extensive water vapor exospheres.

The bright infrared source IRC +1°216 is one well-studied example of what the future Solar System might look like. This particular proto-planetary nebula has been resolved into a series of high-density concentric shells caused by the thermal pulsing and episodic mass-loss of the central Mira variable star CW Leonis (Figure 4.6). The central star is truly a giant, with a radius estimated to be some 500 times larger than the size of the Sun (it would fill our Solar System out to the orbit of Jupiter), and a luminosity some 7,000 times greater. The CW Leonis system also supports a large surrounding cloud of water vapor. Studied in detail with the Submillimeter Wave Astronomy Satellite (SWAS), it is estimated that the water vapor cloud is composed of some four Earth masses of sublimated ice. The question, of course, is where did all this ice come from? The answer is thought to be from billions of comet-like bodies. Just as in our Solar System, it is believed that CW

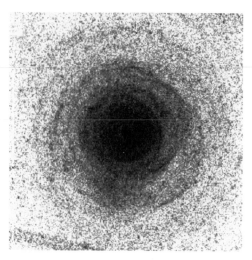

FIGURE 4.6. Multiple-shell structure in the extended envelope surrounding the carbon star CW Leonis (IRC +1°216). The system is some 150 pc distant, and the outermost ring has a diameter of about 0.1 pc. The rings are believed to relate to variations in the mass-loss rate of CW Leonis. (Image courtesy of Dr. Nicolas Mauron)

Leonis has a surrounding swarm of large cometary bodies—the equivalent of our Kuiper Belt and Oort Cloud – and it is these objects that are in the process of evaporating *en mass* due to the high luminosity of the central star.

During the Sun's planetary nebula phase it is likely that Earth will be stripped of its rocky mantle to reveal its metal core—a compact nickel- and iron-rich remnant. This pared-down Earth will eventually find itself in a close orbit about the white dwarf Sun. What happens next partly depends upon how strong the magnetic field of the white dwarf Sun is. All white dwarfs are observed to have magnetic fields, but some show field strengths as high as tens of mega-Gauss.[26] Under these latter conditions it is possible that an electric current loop can form between the central white dwarf and the stripped-down planetary-core. This is essentially a scaled-up version of the interaction between Jupiter and its moon Io in our Solar System. Looking at the consequences of a white dwarf-conducting planetary core current loop forming, Jianke Li and co-workers[27] have suggested that the atmosphere of the white dwarf star will become heated in the regions close to its magnetic poles (Figure 4.7). This extra heating, it is then suggested, will

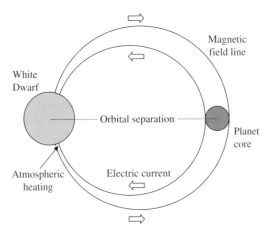

FIGURE 4.7. White dwarf planetary core system. The electric current is generated by the conducting planet moving through the white dwarf's magnetic field lines (a consequence of Ampere's law). As shown here, the orbit of the planetary core is perpendicular to the plane of the page. When a closed circuit has been formed, both the white dwarf's atmosphere and the planet are heated. (Diagram based upon Figure 1 of Note 26.)

result in observable effects. Indeed, Li and colleagues argue that the observed optical emission from the white dwarf star GD356 might be explained by the presence of a small planetary core companion. In this way, the white dwarf star GD356 might just be a model for Earth's distant future if no asteroengineering of the Sun is performed.

What about the planets in the outer Solar System? Although the atmospheres of Venus and Earth will be stripped by the Sun's enhanced red-giant wind, it is likely that Jupiter and Saturn will survive mostly intact. Evidence for this possibility was recently discovered by Peter Maxted (University of Keele, in the UK) and co-workers when they studied the white dwarf star WD0137-349. Surprisingly, they found that the white dwarf had a close binary companion. At a distance comparable to that separating Earth and the Moon, the white dwarf is being orbited every two hours by a brown dwarf companion.[28] The mass of the brown dwarf is estimated to be about 55 times greater than that of Jupiter, but the fact that it survived the red-giant and the planetary nebula phase of its white dwarf host is remarkable, and it suggests that giant planets (such as Jupiter in our Solar System) are able to withstand the ravages imposed during the end-phase evolution of their parent stars. As a consequence of losing orbital energy through gravitational wave generation, the brown dwarf in the WD0137–349 system will have a much smaller orbit. Indeed, in the deep future, about 1.5 billion years from now, once the orbital period of the brown dwarf drops to about one hour, then matter transfer to the white dwarf will commence, leading to the formation of a so-called cataclysmic variable.[28]

A Moving Imperative

It was suggested in Chapter 1 that a number of ancient extraterrestrial civilizations may have already experienced the aging of their parent star to the red-giant phase. The number of such civilizations affected since our galaxy formed can be estimated according to the rate at which interstellar matter is being converted into stars per unit time. It is generally taken by astronomers that the current star formation rate SFR(t) within our galaxy amounts to something like 2 $M_\odot$ of interstellar material being 'converted' into actual stars per year. However, since most stars are less massive than the Sun, this SFR translates into something like one actual star with a mass greater than or equal to the Sun being formed each and every year. In the following calculation it will be assumed that the star formation rate has been constant[29] since the galaxy formed and, consequently for us, SFR(t) = constant. Now it is only the stars born after the initial formation of our galaxy that we are interested in, since these are the stars that will have enhanced heavy element abundances. (Recall from Chapter 2 that it is the supernova endphases of massive stars that have very short lifetimes and produce the heavy elements beyond hydrogen and helium that are vital for the formation of planets and living entities.) The gas clouds out of which the oldest stars within our galaxy formed were essentially composed of pure hydrogen and helium.

The birthrate function BRF(M, t), which accounts for the total number of stars of mass M that have formed in a given time t, is written in terms of the SFR(t) and the so-called initial mass function IMF(M), such that BRF(M, t) = IMF(M) x SFR(t). The IMF is an expression that describes how the mass distribution of stars is divided. The IMF is actually a complicated function of stellar mass, but for stars with masses similar to that of the Sun it can be expressed as a power law with $IMF(M) \sim M^{-2.35}$. This indicates that, in general, there are numerically more low-mass stars than massive ones. Figure 4.8 illustrates the variation in the total number of stars formed per year within two specific mass ranges. The total number of stars formed, within the specified mass ranges, over a time interval t will be equal to the areas under the lines shown in the figure (technically this is equivalent to integrating the birthrate function over a given time interval and mass range). Accordingly, the total number of stars formed over the past 12 billion years in the mass range 0.5 to 1.3 $M_\odot$, corresponding to those stars that might possibly support advanced life bearing planets, is $N(0.5 \Rightarrow 1.3) \approx 2.2 \times 10^{10}$ stars. At this stage, however, what we would like to know is how many of these 22 billion stars have evolved off of the main-sequence to become red-giants. To estimate this number we first need to find the mass of a star that has a main-sequence lifetime equal to that of the age of the galaxy: $\tau_{ms}(M) = t(now) = 12 \times 10^9$ years. The main-sequence lifetime of a star[30] can be expressed purely in terms of its mass, and a star of mass M solar masses has a main-sequence lifetime $\tau_{ms} \approx 10^{10}/M^3$ years. In this way it turns out that, our galaxy formed the lowest mass star that

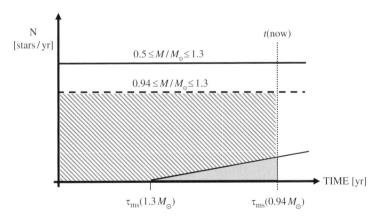

FIGURE 4.8. Schematic variation of the number of stars formed (within a given mass range) against time. The area under the solid line indicates the total number N of stars formed in the mass range from 0.5 to 1.3 $M_\odot$. The solid gray shaded area corresponds to those stars that have evolved off the main-sequence to become red-giants. The cross-hatched area corresponds to those stars still on the main-sequence.

could have evolved off the main-sequence to become a red-giant has a mass of 0.94 $M_\odot$. Again, referring to Figure 4.8, the number of stars born in the mass range 0.94 to 1.3 $M_\odot$ is schematically given by the area under the dashed line. A detailed calculation indicates that the number of stars formed over the past 12 billion years with masses between 0.94 and 1.3 $M_\odot$ is N(0.94 $\Rightarrow$ 1.3) $\approx$ 4.6 x 10^9. Remember, these are the stars that will have enhanced heavy element abundances and may harbor life-bearing planets.

To finish our calculation we now need to determine how many of the N(0.94 $\Rightarrow$ 1.3) $\approx$ 4.6 x 10^9 stars are still on the main-sequence. The main-sequence lifetime of a 1.3 $M_\odot$ star is about 4.5 billion years, so clearly all of the stars that formed with this mass (and greater) shortly after the galaxy itself formed will no longer be main-sequence stars. Indeed, after a time of, say, 6 billion years these stars will have become white dwarfs. Any 1.3 $M_\odot$ star formed within 4.5 billion years of the present, however, will still be on the main-sequence. Again, a detailed calculation reveals that the number of stars in the mass range 0.94 to 1.3 $M_\odot$ that have formed during the past 12 billion years and are still in their main-sequence phase at the present time is N_{MS}(0.94 $\Rightarrow$ 1.3) $\approx$ 3.1 x 10^9 stars. Finally, therefore, the number of stars in the mass range 0.94 to 1.3 $M_\odot$ that have become red-giants since the galaxy formed is N_{RG}(0.94 $\Rightarrow$ 1.3) = N(0.94 $\Rightarrow$ 1.3) - N_{MS}(0.94 $\Rightarrow$ 1.3) = 1.5 x 10^9 stars. (This corresponds to the gray shaded area in Figure 4.8.)

Not all of the 1.5 billion stars with masses between 0.94 and 1.3 $M_\odot$ that have become red-giants since the galaxy formed will have had accompanying planets. Modern-day observations[31] suggest, however, that at least 15 percent of Sun-like stars have planets with orbital radii of less than 5 AU (the orbital radius of Jupiter in our Solar System). Although the observations also indicate that the occurrence of planets is strongly correlated with the heavy element abundance of the parent star (indicating that the more recently formed stars have a higher probability of harboring planets), it would seem that of order 200 million stars with potential planetary systems will have become red-giants since the galaxy formed. It is not known, of course, whether any of these 200 million systems had planets situated within the habitability zone and whether intelligent life ever evolved. Indeed, we are back at the Drake equation [Equation (1.1)] problem discussed in Chapter 1. This being said, Professor Ben Zuckerman, whose original arguments we have essentially followed above,[32] suggests that it is likely that somewhere between 1 and 1,000 civilizations will have faced the consequences of their parent star becoming a red-giant. On this basis he also suggests that the galaxy should be "saturated with extraterrestrial creatures."

The existence of Fermi's Paradox argues that, in spite of Zuckerman's exuberance, not one of the 200 million stars that might have sustained an advanced civilization, but have now evolved into red-giants, has produced a race that has successfully colonized the galaxy (or else, as the paradox states, they should now be here in our Solar System). This observational situation suggests a number of possible scenarios:

1. Civilizations simply die when their parent star becomes a red-giant.
2. No civilization has ever survived long enough to worry about its parent star becoming a red-giant.
3. Interstellar space travel is not a viable means of survival and escape from a planetary system once its parent star becomes a red-giant.
4. The adoption of star-engineering and stellar rejuvenation processes have negated the red-giant evolution imperative for advanced civilizations to move away from their home worlds.

Within the context of the ideas being discussed in this book, it is the fourth scenario that is of particular interest, which is the topic of the next chapter.

Notes and References

1. Liquid water cannot exist upon the surface of Mars now because of the low surface pressure provided by its atmosphere. If liquid water were released upon the Martian surface it would rapidly freeze and then sublimate into a gas. Liquid water could have existed on the surface of Mars in the past since its atmosphere was more substantive then. It is possible –and highly likely – that subsurface liquid water exists on Mars to

this very day. Indeed, Michael Malin and co-workers [Present-day impact cratering rate and contemporary gully activity on Mars, *Science*, **314**, 1573–1577 (2006)] argue that images obtained with the Mars Global Surveyor satellite indicate that new gully deposits have formed within at least two craters situated in the Centauri Montes and Terra Sirenum regions of Mars. The deposits were certainly formed within the last seven years and are interpreted as being due to the flow of liquid water.

2. Equation (4.1) is derived according to the assumption that Earth is an approximate blackbody radiator. At a distance d from the Sun the energy received at the surface of a planet per second will be $E_{received} = \sigma_P (1 - A) L_\odot / (4 \pi d^2)$, where σ_P = is the planet's cross-section area, $\sigma_P = \pi R_P^2$, and A is the atmospheric albedo. The amount of energy re-radiated by the planet back into space because it is a blackbody radiator of temperature T_P will be given by the Stefan-Boltzmann law, such that $E_{radiated} = \varepsilon 4 \pi R_P^2 \sigma T_P^4$, where σ is the Stefan-Boltzmann constant and ε is the emissivity. The equilibrium temperature for the planet, as given by Equation (4.1), is determined under the condition that $E_{radiated} = E_{received}$. It should be noted at this stage that the temperature of the planet is not dependent upon how large it is. The additional term T_{GHE} introduced into equation (4.1) accounts for warming due to the so-called greenhouse effect of an atmosphere. The amount of greenhouse warming will depend upon the composition (specifically, the CO_2 content) and temperature of the atmosphere. The greenhouse warming for Earth at the present time amounts to $T_{GHE} \approx 25$ K.

3. The albedo (A) accounts for how much of the Sun's energy is reflected back into space before it can heat the planet's surface. The emissivity (ε) accounts for how efficiently the planet radiates its absorbed energy back into space. In general the albedo and emissivity will vary with temperature, atmospheric and planetary surface composition, and the wavelength of the incident and re-radiated radiation. For a perfect blackbody radiator, $A = 0$ and $\varepsilon = 1$.

4. We note that $L(t = 0) < L(t = 4.5$ billion years$)$ [see Table 3.2], and this suggests that solar heating alone would not have been sufficient to stop Earth's initial oceans from freezing. Since it

is (arguably) evident that the oceans didn't freeze, this suggests that there was an additional heating term. It is generally argued that the extra heating was due to a higher greenhouse heating term T_{GHE} in the distant past, when Earth had a richer CO_2 atmosphere. For the young Earth, global warming was a good thing. For the current Earth it is a worrying phenomenon, since the Sun is about 45 percent more luminous now than when it first settled onto the main-sequence 4.5 billion years ago.

5. See, for example, the detailed model calculations presented by James Kasting and co-workers in Habitable zones around main-sequence stars. *Icarus*, **101**, 108–128 (1993). A simplified time-dependent Earth climate model is considered by Ken Caldeira and James Kasting in, The life span of the biosphere revisited. *Nature*, **360**, 721–723 (1992).

6. The physicist Steven Toulmin once remarked that definitions are like belts. The shorter they are, the more elastic they need to be. Although the definition for the habitability zone is not short, it does warrant a few addendums. One case in point has been described by David Stevenson [Life-sustaining planets in interstellar space? *Nature*, **400**, 32 (1999)]. Stevenson points out that planet formation is a highly dynamic process, and it is conceivable that Earth-mass planets are formed and then ejected from a bound orbit into interstellar space. At first thought this suggests that the planet is doomed and that any atmosphere and/or surface water will rapidly freeze out. The situation, however, is more complex and subtle. Stevenson argues that the slow release of internal heat energy, built up by the accretion process and the decay of radioactive elements, can sustain a liquid layer on an Earth-mass planet for perhaps several billion of years, even if it is situated in the cold depths of interstellar space. Perhaps – even on these exotic, perma-nently dark worlds – elementary life can evolve and may be even prosper for a short while.

7. B. W. Jones, P. N. Sleep, and J. E. Chambers. The stability of the orbits of terrestrial planets in the habitable zones of 254–262 (2001). The star ρ CrB is slightly less massive than the Sun ($M = 0.95\ M_\odot$) and is estimated to be about 6 billion years old (some 1.5 billion years older than the Sun). The star 47 UMa is

slightly more massive than the Sun ($M = 1.03\ M_\odot$) and nearly twice as old, with an estimated age of 7 billion years.

8. D. M. Williams, J. F. Kasting, and R. A. Wade. Habitable moons around extrasolar giant planets. *Nature*, **385**, 234–235 (1997). R. C. Domingos and co-workers [Stable satellites around extrasolar giant planets. *Monthly Notices of the Royal Astronomical Society*, **373**, 1,227–1,234 (2006)] derive analytic expressions for the semi-major axis and eccentricity of stable satellite orbits within exoplanetary systems.

9. Christopher Lovis et al., An extrasolar planetary system with three Neptune-mass planets. *Nature*, 441, 305–309 (2006).

10. The structure and properties of the four largest moons of Jupiter are described in A. Showman and R. Malhotra, The Galilean satellites, *Science*, **286**, 77 (1999). A number of models describing the possible internal structure of Europa are presented in J. Anderson et al., Europa's differentiated internal structure: inferences from two *Galileo* encounters, *Science*, **276**, 1,236 (1997). The surface structure of Europa is described by F. Nimmo and co-workers in Europa's icy shell: past and present state, and future exploration, *Icarus*, **177**, 293 (2005).

11. Alon Retter et al., The planets capture model of V838 Monocerotis: conclusions for the penetration depth of the planet(s). *Monthly Notices of the Royal Astronomical Society*, **370**, 1,537–1,580, 2006.

12. The key effect that has to be considered at this stage is that of greenhouse warming. As the Sun's temperature increases so the evaporation rate of Earth's oceans also increases, and a moist greenhouse effect will develop in which a dense water vapor-laden atmosphere overrides a near boiling ocean. The next phase relates to timescale; if the Sun reaches a luminosity of 40 percent brighter than it is now and the oceans have not fully evaporated, then a runaway greenhouse effect comes into play, trapping heat near to Earth's surface and pushing the temperature to many hundreds of Kelvins. Earth's surface may actually melt in some places under the runaway greenhouse scenario.

13. Extremophiles are microorganisms that can survive and thrive, under conditions that would be fatal to most other organisms. The thermophiles, and hyperthermophiles, for example, are

found in environments where the temperature varies from 50 to 80°C and 80 to 110°C, respectively. These microorganisms thrive, for example, in the deep ocean floor environments surrounding hydrothermal vents, or 'black smokers.' Other microorganisms such as the psychropiles can tolerate extreme cold, while the halophiles thrive under high salinity conditions. Peter Ward and Donald Brownlee have attempted to describe the final stages of life on Earth in their interesting, but unnecessarily doom-laden book, *The Life and Death of Planet Earth* [Owl Books, New York (2002)].

14. Technically T_{decay} provides what is called the *e*-folding time, which is the characteristic time over which the orbital radius changes by a factor of $e = 2.7183$. The complete spiral-in destruction time for Mercury would probably correspond to a few *e*-folding times.

15. The envelope density is estimated from the red-giant models computed by William Rose and Richard Smith [Final evolution of a low-mass star II, *Astrophysical Journal*, **173**, 385–391 (1972)]. We use the model corresponding to a star having a radius of 164 $R_\odot$ and luminosity of 2500 $L_\odot$. This model atmosphere corresponds to the Sun at the red-giant tip (point 4 in Figure 3.10).

16. A similar calculation to the one presented here for Mercury was made for Earth by Samuel Vila [Survival of Earth and the future evolution of the Sun, *Earth, Moon and Planets*, **31**, 313–315 (1984)]. Vila finds that if the envelope of the red-giant Sun does extend to encompass Earth's orbit, then the *e*-folding time for orbital decay is about 5,000 years. Again, this is a very short timescale, and destruction of Earth is assured. Goldstein [The fate of Earth in the red-giant envelope of the Sun, *Astronomy and Astrophysics*, **178**, 283–285 (1987)] has presented a more detailed calculation for Earth's orbital decay time – including gas drag forces – and finds an *e*-folding time of just a few hundred years—a timescale even more rapid than that found by Vila. It appears that a planet is rapidly destroyed once it begins to encounter the gas envelope of its red-giant parent star.

17. The orbital angular momentum h of a mass m moving with velocity V in a circular orbit of radius a is $h = a\,m\,V$. If this is

combined with Kepler's third law, then it turns out that, $a M = (h/2 \pi m)^2$ = constant, where M is the mass of the object about which the smaller mass m is moving around. The key point about angular momentum is that it is a conserved quantity, meaning that $h_{final} = h_{initial}$. In this manner the product $a(t) M(t) = a(0) M(0)$ = constant, where t is the time, must hold true.

18. Peter Schroder, Robert Smith, and Kevin Apps [Solar evolution and the distant future of the Earth, *Astronomy and Geophysics*, **42**, 6.26–6.29 (2001)] argue that recent observations indicate modest mass-loss rates during the red-giant phase of solar mass stars. Consequently their solar model expands to destroy Venus. The solar evolutionary calculations published by I-J. Sackmann and co-workers [Our Sun III: present and future, *The Astrophysical Journal*, **418**, 457–468 (1993)], on the other hand, assume a relatively high mass-loss rate during the Sun's red-giant branch phase and, consequently, Venus survives, as shown in Figure 4.3.

19. Alain Léger et al., A new family of planets? Ocean-planets. *Icarus*, **169**, 499–504 (2004). Interestingly, Leger and collaborators point out that the low density of ocean planets dictates that they should be relatively large and, hence, good potential targets for spectroscopic study. This, in turn, suggests that, should they be found, they are attractive candidates for surveys looking for bio-signatures—such as ozone, or O_3. Indeed, the *Terrestrial Planet Finder* (TPF) mission, currently under development by NASA and due for launch circa 2015, is being designed to specifically look for life-related signatures in nearby extrasolar planetary systems.

20. Don Korycansky, G. Laughlin and F. Adams, Astronomical engineering: a strategy for modifying planetary orbits. *Astrophysics and Space Science*, **275**, 349–366 (2001). See also D. Korycansky, Astroengineering, or how to save Earth in only one billion years. *Reviews of the Mexican Astronomical Association*, **22**, 117–120 (2004).

21. With reference to Equation (4.1), the Korycansky scenario requires that the $L(t)/d^2$ term remains constant with time. The Sun's luminosity on the main-sequence varies approximately as: $L(t) = L_\odot/(1 - 0.38 \, t/\tau)$, where $\tau = 4.55 \times 10^9$ years, and where t is expressed in years from the present time. (This formula

is taken from the Caldeira and Kasting paper introduced in Note 5.). With $L(t)$ specified the required increase in the size of Earth's orbital radius $d(t)$ can be determined.

22. The yearly weather cycle (winter, spring, summer, and autumn) is driven primarily by the tilt of Earth's spin axis to the ecliptic (that is, Earth's orbital plane). This angle, which amounts to some 23.5°, is called the *obliquity of the ecliptic*. Any change in Earth's obliquity will result in distinct climate changes, and this is where the Moon comes in as a stabilizing agent. A numerical simulation carried out by J. Laskar and co-workers [Stabilization of the Earth's obliquity by the Moon. *Nature*, **361**, 615–617 (1993)], for example, indicates that the Earth's obliquity would vary chaotically and dramatically over many degrees if the Moon did not exist. On the other hand, Darren Williams and co-workers [Low-latitude glaciation and rapid changes in the Earth's obliquity explained by obliquity-oblateness feedback. *Nature*, **396**, 453–455 (1998)] argue that the climate itself can modulate Earth's obliquity. They reason, for example, that the buildup of massive ice sheets during past glacial cycles has actually reduced Earth's obliquity.

23. Colin McInnes, Astronomical engineering revisited: planetary orbit modification using solar radiation pressure. *Astrophysics and Space Science*, **282**, 765–772 (2002).

24. Leonid Shkadov, Possibility of controlling Solar System motion in the Galaxy. 38th *IAF, International Astronautical Congress*, Brighton (1987). Paper 1AA-87–613.

25. Luc, F. A. Arnold, Transit light-curve signatures of artificial objects. *Astrophysical Journal*, **627**, 534–539 (2005). On the basis that an artificial transit sail is constructed in the main-belt asteroid region of our Solar System, the sail would then have to be maneuvered closer in toward the Sun so that transits would both repeat frequently and be visible to a large potential audience.

26. Earth's magnetic field strength is about 0.5 Gauss, while that of the Sun's is 1 Gauss. Jupiter supports the strongest magnetic field in the entire Solar System, with field strength of some 8 Gauss. Indeed, decameter radio emission bursts are regularly recorded from Jupiter, and these bursts are synchronized with the orbital period of its moon, Io. It is the motion of Io through

Jupiter's magnetic field that produces the so-called synchrotron radio emissions.

27. Jianke Li, L. Ferrario, and D. Wickramasinghe. Planets around white dwarfs, *The Astrophysical Journal*, **503**, L151–L154 (1998).

28. P. F. L. Maxted et al., Survival of a brown dwarf after engulfment by a red-giant star. *Nature*, **442**, 543–545 (2006). White dwarfs rarely have brown dwarf companions – such pairings occurring in less than 0.5 percent of the systems containing white dwarfs. The WD–0137-349 system will eventually form into a short-period cataclysmic variable, with matter being transferred from the brown dwarf to the white dwarf. A full blown Type I supernova (see Figure 2.13 and Note 13, Chapter 2) will not occur since there isn't enough mass in the brown dwarf to push the white dwarf beyond the Chandrasekhar limiting mass. Hydrogen-rich matter will accumulate on the white dwarf's surface, however, and this will periodically undergo runaway thermonuclear reactions to produce a nova-like outburst.

29. The star formation rate in our galaxy has not been constant during the past 12 billion years. It is usually assumed that the SFR decreases exponentially with time, but for our purposes a constant rate will produce the order of magnitude result being sought.

30. Here we have combined the main-sequence lifetime Equation (2.1) with the mass-luminosity Equation (3.12). We have assumed, however, that for solar mass stars the luminosity varies as the mass to the fourth power. The important point about the main-sequence lifetime is that it is shorter for more massive stars. Although the Sun has a main-sequence lifetime of about 10 billion years, a 10 solar mass star has a main-sequence lifetime of about 10 million years. So, although massive stars do have more hydrogen available for consumption, they use it up (in the sense of radiating energy into space) more rapidly.

31. A good technical review is provided by G. Marcy et al., Observed properties of exoplanets: masses, orbits and metallicities. *Progress of Theoretical Physics Supplement*, No. 158, 1–19 (2005). The percentage of Sun-like stars harboring planets

within 5 to 6 AU of their parent stars may even be as high as 25 percent. This higher percentage results in there being something like 400 million stars with masses between 0.93 and 1.5 $M_\odot$ harboring planets and having left the main-sequence since the Milky Way galaxy formed. A more general review of the properties of exoplanets is given in the April 2007 issue of *Astronomy Now* Magazine.

32. B. Zuckerman, Stellar evolution: motivation for mass interstellar migration. *Quarterly Journal of the Royal Astronomical Society*, **26**, 56–59 (1985).

5. Rejuvenating the Sun

In Chapter 3 the physical processes underlying the workings of a Sun-like star were described. In this chapter we will examine the ways in which the properties of a star might, at least in principle, be manipulated by our distant descendants. Specifically, our task is to see how the Sun might be 'engineered' or 'rejuvenated' to enable the continued survival of life on the innermost planets, on timescales greater than the canonical main-sequence lifetime [T > T_{MS} (canonical)]. In the case of Venus and Mars, of course, this clearly means future human life on terraformed worlds. As already stated, the task of the would-be asteroengineer is to find ways to stop the Sun from becoming over-luminous, and from becoming a bloated red-giant – the dire consequences of these effects for the Solar System having been discussed in the last chapter. It turns out, fortuitously for humankind, that these goals are compatible; by stopping the red-giant Sun from coming about, the long-term temperature stability of the inner planets is also maintained.

Perhaps it should be reiterated at this stage that we are not describing in this book exactly how the mechanical part of star engineering can be done. We do not know, for example, what kinds of materials should be used or how to construct the various machines and devices that will be described in this chapter. What we will outline, however, is how the future properties of the Sun might be controlled in principle.

The Engineering Options

As highlighted in Chapter 4, the most important problem that the future star engineer will need to address is that of the Sun's increasing luminosity. Its increase in radius is not so great an issue if we are only concerned with the survival of Planet Earth, but it seems an incredible waste of resources to simply let Mercury and Venus be consumed by an expanding Sun. Equation (4.1) contains the key terms of interest and, indeed, it indicates that for a fixed planetary distance d, the surface temperature of the planet increases as the Sun's luminosity to the one-quarter power—that is as $L^{1/4}$.

If all of the other terms on the right-hand side of Equation (4.1) remain the same, then the surface temperature of any given planet increases by about one degree for every 1 percent increase in the Sun's luminosity. So, to stop Earth from overheating, the star engineer must control the growth of the Sun's luminosity. Indeed, the aim will be to keep the Sun at or at least near its present energy output per unit time. In fact, a slightly less luminous Sun might be desirable. This latter dictate builds upon the suggestion by Professor James Lovelock that the recent glacial – interglacial cycling that has dominated the Pleistocene era is a Gaian response to the enhanced warming of Earth in recent times.[1] The Sun's ideal luminosity was achieved, according to Lovelock, some 2 billion years ago, when it was 15 percent less luminous than now.

How then might the star engineer proceed? The double goal of eliminating the red-giant phase and reducing the Sun's luminosity—the basic act of rejuvenation—can be achieved by manipulating both internal and external quantities. By external quantities we specifically mean the mass of the Sun, and by internal we mean the radial variation in its composition. No one process of manipulation is going to achieve both of the stated goals, so a combination of alteration mechanisms will be required.

Mixing and Mass Loss

In this section we will build upon the results leading to Equation (3.12). Specifically the mass-luminosity relationship indicates that if the Sun is to have the same luminosity at the beginning and the end of its main-sequence phase, then its mass at the end of the main-sequence must be reduced to[2] $\sim 0.3 \, M_\odot$. In other words, the Sun must be slimmed down by some $0.7 \, M_\odot$ worth of material.

The example considered above assumes that the Sun has a homogeneous composition. Detailed numerical models, however, have shown that even if a star has an inhomogeneous composition (where the envelope, for example, is more hydrogen-rich than the core), the evolution with mass-loss is always at a lower luminosity. Figure 5.1 illustrates, in a schematic way, the effects of mixing and mass-loss on the evolution of a star.[3] It can be seen from the figure that the effect of inducing greater and greater amounts of additional mixing within the interior of a star results in the red-giant phase being killed off. Rather than evolving into a low temperature, large red-giant at core hydrogen exhaustion, a fully mixed star evolves into a luminous, slightly larger, and higher temperature star. To the star engineer this result illustrates how the bloated red-giant stage of the Sun can be avoided and, accordingly, methods of mixing the Sun's interior will have to be developed. The evolution of a fully mixed star with mass-loss is again toward higher temperatures, but now the mass-loss

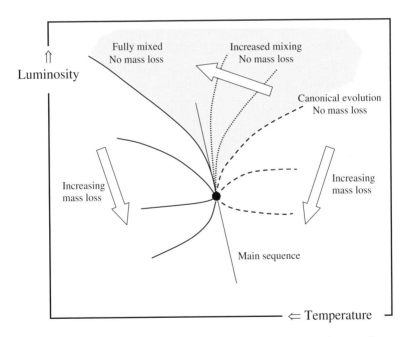

FIGURE 5.1. Schematic diagram showing the effects of mass-loss and mixing on the evolution of a star.[3] The standard non-mixed evolutionary tracks are shown as dashed lines. The chemically homogenous evolutionary tracks are shown as solid lines. The effect of partially mixing a star is shown by the dotted lines. The shaded region indicates the effect of mixing without mass-loss.

results in lower luminosities being achieved—the greater the mass-loss, the lower the luminosity for any given composition. If the mass-loss is very high, the evolution can proceed to values lower than the initial main-sequence luminosity. A non-homogeneous star evolving with mass-loss is also, for a given composition, less luminous than the non-homogeneous zero mass-loss model. The evolution is still towards lower surface temperatures, however, and unless extreme amounts of mass are removed from the star, the red-giant phase will still occur.

To sum up so far, for the Sun to avoid its bloated red-giant phase, and for it to evolve at near constant luminosity, both mass-loss and the (near) complete internal mixing of its chemical elements must be engineered.

Adding to the Pressure

The extent of a star's hydrogen-burning phase (its main-sequence lifetime) is expressed, constants aside, by Equation (2.1) as $T_{MS} \sim M/L$. For a star of fixed mass M this relationship indicates that the hydrogen-burning phase might be extended if the star can be made to operate at a lower luminosity L. Such a possibility exists, provided that the would-be star engineer can find a way of introducing some non-thermal pressure support to help the star remain in hydrostatic equilibrium at a lower temperature. In this manner the pressure $P(r)$ at any point within a star is composed of two terms: the gas pressure P_{gas} and a non-thermal pressure, P_{NT}. Two examples of non-thermal pressure support relate to strong magnetic fields and rapid internal rotation. If the pressure is written as $P(r) = P_{gas} + P_{NT} = P_{gas}(1 + \delta)$, where $\delta = P_{NT}/P_{gas}$ is the ratio of the non-thermal to the thermal pressure support, then Equation (3.5) for the central pressure can be re-derived, and accordingly[4] $T_C = [G\, m_H/3\, k]\, \mu(1 - \delta)\, M/R$. When $\delta = 0$ there is no non-thermal pressure support, and we recover Equation (3.5). As δ increases above zero, however, the central temperature T_C required to achieve hydrostatic equilibrium is reduced.

The reduced central temperature that results from the introduction of an additional non-thermal pressure term dictates that the PP chain will run less efficiently [recall Equation (3.11)] and, consequently, the luminosity will be reduced. Indeed, the mass-luminosity relationship for Sun-like stars [Equation (3.12)] becomes $L = L_{KR}[\mu(1 - \delta)]^{7.5} M^5/(1 + X)$. Now we recover the result that as δ increases from zero so, for a given composition and fixed stellar mass, the luminosity is reduced. If we go back to the expression for the main-sequence lifetime, $T_{MS} \sim M/L$, then the effect of introducing additional non-thermal pressure support is to increase the main-sequence lifetime to $T_{ntps} = T_{MS} (1 - \delta)^{-7.5}$. Table 5.1 indicates the effect of introducing increasing amounts of non-thermal pressure support. The greater the value of δ, the longer the main-sequence lifetime and the lower the luminosity of the star at core hydrogen exhaustion. Indeed, as Table 5.1 indicates, if the non-thermal pressure support in the Sun could be increased to a value of order 10 percent, then its main-sequence lifetime would be increased by a factor of two, to of order 20 billion years, and its luminosity at core hydrogen exhaustion would be reduced by a factor of about one-half compared to its canonical (and fully mixed) evolutionary value at core hydrogen exhaustion.

To drive the central temperature of the Sun down by engineering additional non-thermal pressure support would be far from simple—at least to begin with. The Sun's magnetic field is currently generated within its convective outer envelope, and there is no straightforward way to sustain a central magnetic field. By inducing mass-loss, however, (as indicated in Figure 3.6) the rejuvenated Sun will develop a deeper and deeper outer convection zone as its mass is physically reduced. Once the mass is below about 0.5 $M_\odot$, then the modified Sun will be nearly fully convective, and the enhanced magnetic dynamo action that will then operate should

Table 5.1. The effect of additional non-thermal pressure support upon the main-sequence lifetime. The first column indicates the value of δ. Columns 2 and 3 show the resultant increase in the main-sequence lifetime and the reduced luminosity (when $X = 0$) for non-zero values of δ.

δ	T_{ntps}/T_{MS}	$L(X=0, \delta)/L(X=0, \delta = 0)$
0.01	1.1	0.91
0.05	1.5	0.67
0.1	2.2	0.45
0.2	5.3	0.19
0.4	46.1	0.02
0.5	181.0	0.005

result in the generation of a significant non-thermal pressure effect at the center. If this late-stage rejuvenation process can be made efficient, then the mass-loss rate could be modified downward, since the non-thermal pressure support provided by a strong central magnetic field will cause a reduction in the Sun's luminosity.

Although the non-thermal pressure support due to magnetic fields might well become important during the later stages of the Sun's rejuvenation (once its mass is reduced below about 0.5 $M_\odot$), rotation-related non-thermal pressure support might conceivably be induced during the early rejuvenation stages. In this case the Sun's internal gravity is counteracted by the centrifugal repulsion that comes about because of the rotation. The faster the Sun can be made to rotate the greater the rotational non-thermal pressure support. In addition, detailed numerical modeling indicates that enhanced rotation within a star leads to more extensive mixing of the chemical elements and in the enhancement of the overall magnetic field. Both of these results are desirable star-engineering byproducts. As we will see in more detail below, David Criswell has described a method by which a star might be spun-up and 'mined' at the same time.

The Opacity Effect

A second method by which a star's luminosity might conceivably be reduced, thereby enhancing its hydrogen-burning lifetime, is through the manipulation of the opacity of stellar material. As described in Chapter 3, the opacity κ is a measure of how much the stellar material hinders the passage of radiation through it. The greater the opacity of a region, the longer it takes the radiation to pass through it. The effect of increasing the opacity is to reduce the mean-free path length $l = 1/(\kappa\rho)$ between photon interactions. This, in turn, results in a longer photon diffusion time T_{PD} [as given by Equation (3.6)]. Examining Equations (3.7) and (3.9) further indicates that an increase in the opacity results in a lower surface temperature and a smaller luminosity. In other words, the greater the opacity of stellar material, the cooler and less luminous a star will become. The lower temperature and luminosity, however, will result in the star becoming larger, in accordance with the

Stefan-Boltzmann law [as given by Equation (3.8)]. In a technical, but very readable, paper on the order-of-magnitude theory of stellar structure,[5] Professor George Greenstein of Amherst College shows that if the opacity is increased by a factor of 100 over that provided by electron scattering, then the luminosity will be reduced by a factor of 100 and the surface temperature will be reduced by a factor of about 10.

A significant fraction of the stellar opacity within Sun-like stars is due to the heavy elements (measured by the Z abundance; see Table 3.1). Simply adding additional heavy element-rich material to the Sun will not actually help in reducing its luminosity, since this process only adds more mass to the Sun, and this will offset the required effect as indicated by the mass-luminosity Equation (3.12). One might envision a dialysis-like process, however, whereby solar helium is extracted and then preferentially replaced with, say, iron or some such similar heavy element. This, however, would be an exceptionally complicated process, and there is no ready-at-hand supply of heavy element material with which to effectively seed the Sun.

The Tools of the Trade

A summary of the basic rejuvenation processes described above are presented in Table 5.2. The processes apply not just to the Sun, but to all stars that are likely to harbor habitable planets and possibly advanced life forms. There are a few additional engineering options that will be introduced below, but for the moment it seems[6] that the best way to begin the process of rejuvenating the Sun, and thereby saving Earth from complete devastation, is to engineer additional mixing within its interior and enhance its mass-loss rate. The non-thermal pressure support and opacity alteration methods may still be employed at a later time, but they might be more advanced processes (both physically and temporally) in the rejuvenation sequence. In short, the solar rejuvenation process will initially proceed by manipulating the physical body of the Sun. Indeed, just like a human being, the Sun can be rejuvenated by slimming down and consuming its food (in the form of hydrogen, of course) more efficiently.

Table 5.2. Summary of the principal processes available for rejuvenating a star.

Process	Reason [key equations]	Response
Mass reduction	Mass-luminosity relationship [(3.12)]	Luminosity reduction
Non-thermal pressure support	Reduced thermal pressure requirement at center results in lower central temperature [(3.5), (3.7), (3.9), (3.11)]	Reduction in central temperature leading to a reduced luminosity and lower surface temperature
Opacity increase	Reduced mean-free path length (*l*) [(3.6), (3.7) and (3.9)]	Reduction in luminosity and surface temperature
Mixing of elements	Avoids a core–envelope composition discontinuity [see Figure 5.1]	Kills off red-giant stage expansion

A Homogeneous Star Model

In Chapter 3 the mass-luminosity relationship for homogeneous Sun-like stars was described [see Equation (3.12)]. Accordingly, the luminosity L is related to the mass M and the chemical compositions as $L = L_0 \, \mu^{7.5} \, (1 + X)^{-1} \, M^5$, where $L_0 = L(X = X_0)$ is a constant, μ is the mean molecular weight [see Equation (3.4)] and $0 \leq X \leq X_0$ is the hydrogen mass fraction of the stellar gas. In Appendix A it is explained that when the mass fraction of the chemical elements other than hydrogen (X) and helium (Y) are small, then the mass-luminosity relationship can be written as $L(X) \approx L_0 \, (1 + X)^{-16} \, M^5$. This modified formula[7] expresses the luminosity of a fully-mixed star of mass M solely in terms of the hydrogen mass fraction X. As a star ages, however, and hydrogen is consumed via the proton-proton chain of fusion reactions, the value of X will decrease from its initial $X_0 = 0.7$ to zero. For a constant mass star, therefore, $L(X = 0)/L(X = X_0) = 1/(1 + X_0)^{-16} \approx 4900$. In other words, for a fully mixed star, the luminosity is expected to increase by a factor of several thousand during its main-sequence (that is hydrogen-burning) lifetime. Detailed numerical models confirm this brightness increases to about the correct order of magnitude[8]. Simply fully mixing a star will not produce the rejuvenation effect being sought, and while the effect will kill off the giant phase it will not stop the Sun from becoming over-luminous.

Introducing Mass-Loss

Mass-loss, literally the reduction in the mass of a star, is an observed phenomenon. The Sun, for example, currently loses mass via the so-called solar wind, while other stars are observed to spew so much material into space that they become surrounded by glowing nebulae. It is the solar wind that is responsible for producing auroral displays in Earth's upper atmosphere and for occasionally interrupting the workings of communications space-craft. The solar wind emanates from the Sun's outer corona and careens past Earth with a speed that varies between 200 and 800 km/s. The mass-loss rate for the Sun amounts to some 1.5×10^{-14} $M_\odot$/yr, equivalent to about a billion kilograms of material being ejected into space per second. The current annual mass-loss rate from the Sun is minuscule compared to its total mass, and even if it continued to lose mass at its current rate for the rest of its main-sequence lifetime, then only $10^{-4} M_\odot$ of material would be lost into space. This, however, is equivalent to the mass of about 53 Earths. Observations indicate that the mass-loss rate varies in proportion to the luminosity of a star, and it is useful to express the mass-loss rate via the expression $\Delta M / \Delta t = N (L/c^2)$, where ΔM is the amount of mass lost by the star in the time interval Δt. N is a numerical parameter that can vary from zero (indicating no mass-loss) to a value as high as several hundred, L is the luminosity, and c is the speed of light. Using this expression for the mass-loss rate it can be shown (see Appendix A) that the mass of star decreases exponentially as $m = \exp[k (x - 1)]$, where $m = M(X)/M(X=X_0)$, $x = (1 + X)/(1 + X_0)$ and k is a mass-loss rate dependent term defined in Equation (A.7). We now have the result that the mass of a star decreases as X decreases from X_0 to zero, and that the larger the value of the mass-loss parameter N, the more rapidly does the mass decrease with X. Introducing the notation $l = L (X)/L(X = X_0)$, the luminosity of a chemically homogeneous, mass-losing star can be expressed as $l = x^{-16} m^5$, and from this formulism we see that a mass-losing star is always going to be under-luminous when compared to a star of similar composition, but not losing mass.[9] It is this under-luminosity condition that the future star engineer will want to exploit. Indeed, with respect to rejuvenating the Sun and extending the habitable lifetime of Earth, inducing additional mass-loss from the Sun is going to be crucial.

The Fate of the Ejected Material

Earth's average global temperature is described by Equation (4.1), and $T_{surface}$ will vary according to how the L/d^2 term changes with time. (Let's assume for the sake of argument that Earth's albedo and emissivity remain constant.) If the orbital distance d remains fixed, then $T_{surface}$ remains constant, provided the Sun's luminosity L is constant. If, on the other hand, the orbital distance d increases with time, then $T_{surface}$ will only remain constant if the Sun's luminosity L increases in order to keep the L/d^2 term fixed. Since the solar rejuvenation process invokes mass-loss to control the Sun's luminosity, two extreme scenarios can be explored, with the ultimate location of the mass removed from the Sun being the controlling factor. Accordingly,

1. If the material removed from the Sun is fully contained within the region interior to Earth's orbit, then Earth's orbital radius will remain fixed, and $T_{surface}$ will remain constant provided the Sun's luminosity L is kept constant.
2. If the material ejected from the Sun is lost into interstellar space (or at least escapes to the outer Solar System), then Earth's orbital radius will increase in accordance with the conservation of angular momentum, and $T_{surface}$ remains constant provided the ratio L/d^2 = constant is maintained. In this scenario the Sun's luminosity is allowed to increase with time.

In Scenario 2 it is the conservation of angular momentum that will determine the Sun-Earth separation $d(t)$. The increase in Earth's orbital radius with decreasing solar mass is expressed as $d(t) = d(0) \exp[-k(x - 1)]$, where the constants k and x are described in Appendix A, and $d(0)$ is Earth's orbital radius at the onset of the rejuvenation process. For a fully mixed star to evolve under the constant luminosity condition (Scenario 1), the mass-loss rate must vary according to the parameter N_1 (given by Equation A.9). For a fully mixed star to evolve with L/d^2 = constant (scenario 2) the mass-loss rate must vary according to the parameter N_2 (given by Equation A.10).

Scenario 2, as stated before, seems to be an unnecessarily wasteful process. Why not use all or most of the material being removed from the Sun to build or power something else? If some of the material being driven from the rejuvenated Sun can be channeled to Jupiter, for example, then a low-mass synthetic star might conceivably be generated. The minimum mass for the initiation of hydrogen conversion as a *bona fide* star is about 0.1 $M_\odot$ (recall Figure 3.15). If the mass of Jupiter was, therefore, increased by a factor of about 100, it could begin to generate sustained energy to power industrial plants on the Jovian moons and within the main-belt asteroid region. This being said, by generating a low-mass binary companion to the Sun, one would have to be exceptionally careful not to catastrophically alter the orbital dynamics of the planets and asteroids within the inner Solar System. The devil, as always, is in the details.

An Outline Scenario for Rejuvenating the Sun

In this section we will consider the requirements for rejuvenating the Sun under the two mass-loss scenarios described above. For the sake of convenience let's assume that the Sun has been successfully homogenized and that the initial hydrogen mass fraction is $X_0 = 0.7$. Table 5.3 shows the variation in the mass-loss rate as well as the rejuvenated Sun's mass and surface temperature,[10] as a function of X for Scenario 1, where it is just the luminosity that remains constant.

Table 5.4 indicates the variation in the mass, luminosity, and surface temperature of the rejuvenated Sun under mass-loss Scenario 2. Since in this situation it is the ratio L/d^2 that remains constant, Earth's orbital radius d(AU) is also shown in the table.

Table 5.3. Rejuvenated Sun evolution under mass-loss Scenario 1, where $L(X)$ = constant. See Note 10 for an outline of how the Sun's temperature evolution was determined. The mass-loss parameter N_1 is described in Appendix A.

X	N_1	$M(X)/M_\odot$	T (K)
0.7	269	1.00	5780
0.6	286	0.82	5820
0.5	305	0.67	6020
0.4	326	0.53	6270
0.3	352	0.42	6580
0.2	381	0.32	7000
0.1	415	0.24	7680
0.01	453	0.18	9740

The evolutionary tracks corresponding to the two mass-loss scenarios are shown in Figure 5.2, along with a canonical solar evolution track. The rejuvenated Sun's evolutionary path is initially upward, along the main-sequence in the second mass-loss scenario, where L/d^2 is kept constant. In mass-loss Scenario 1, where the luminosity is kept constant, the evolution is simply towards higher surface temperatures with time. The outward expansion of Earth's orbital radius in the second mass-loss scenario (last column of Table 5.4) results in its final location being some 3.4 AU from the Sun. As required by the conservation of angular momentum the orbits of the other planets and the main-belt asteroids will also increase by a factor of about 3.4 over their current values.

Table 5.4. Rejuvenated Sun evolution under mass-loss Scenario 2, in which L/d^2 = constant. See Note 10 for an overview of how the Sun's temperature evolution was determined. The mass-loss parameter N_2 is described in Appendix A.

X	N_2	$M(X)/M_\odot$	$L(X)/L_\odot$	T (K)	d(AU)
0.7	192	1.00	1.00	5780	1.0
0.6	204	0.87	1.3	5980	1.2
0.5	218	0.75	1.8	6200	1.3
0.4	234	0.64	2.5	6460	1.6
0.3	252	0.54	3.5	6800	1.9
0.2	273	0.44	4.4	7240	2.1
0.1	297	0.36	7.7	7970	2.8
0.01	324	0.30	11.4	10,060	3.4

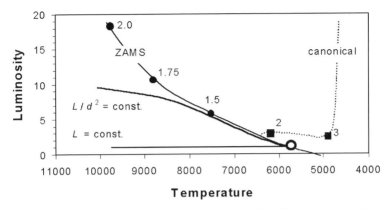

FIGURE 5.2. The Hertzsprung-Russell diagram for the Sun's evolutionary tracks according to various mixing and mass-loss scenarios. The dotted line corresponds to the Sun's standard (canonical) evolutionary track, and the filled squares indicate points 2 and 3 from Table 3.2 (see also Figure 3.10). The filled circles on the main-sequence line indicate the luminosity and temperatures for a range of stellar masses. The large circle indicates the present main-sequence location of the Sun.

The UV Problem

Although it is the Sun's total energy output per unit time (its luminosity) that is important with respect to Earth's climate, the energy radiated by the Sun into space in a specific wavelength region is determined by the Sun's temperature. Indeed, Wien's law tells us that the wavelength (λ_{max}) at which a blackbody radiator emits its greatest energy flux varies with the temperature (T) in such a way that the product $\lambda_{max} T$ is a constant (= 2.8978 x 10^{-3}). For the present Sun, with a temperature $T_{\odot}$= 5780 K, the greatest energy flux is at a wavelength of $\lambda_{max} \approx 501$ nm, which is why the Sun appears to be a yellowish-orange color to our eye. If a star, however, has a temperature twice that of the Sun, then $\lambda_{max} \approx 250$ nm, and this falls in the ultraviolet (UV) part of the electromagnetic spectrum. The increased temperature-related shift of λ_{max} to the ultraviolet for a rejuvenated Sun is a potentially serious problem for future inhabitants of Earth.

The UV segment of the electromagnetic spectrum falls in the wavelength region between 280 and 400 nm. As a result of atmospheric absorption it is mostly UV-A radiation, with 315 < λ

(nm) < 400 that reaches Earth's surface. It is UV-A radiation that enables the skin to generate vitamin D, but more harmfully it is also responsible for producing sunburns and cataracts. The lower flux of UV-B radiation, with $280 < \lambda$ (nm) < 315 that reaches Earth's surface, however, is far more problematic since this radiation can cause direct damage to living cells, resulting in the development of cancerous growths. If the Sun's temperature increases by a factor of two, as is indicated by the rejuvenation processes, its total energy output in the UV part of the electromagnetic spectrum will increase by a factor of about 40. Indeed, the UV-A flux will increase by a factor of 34, while that of UV-B increases by a factor of 66. For the rejuvenation process proceeding with $L(X) = L$ (mass-loss Scenario 1), this is definitely a problem, since the increased UV flux will influence both Earth's atmospheric structure and chemistry in addition to it being potentially deadly to all surface-dwelling life forms.

Shielding Earth from the enhanced UV flux will be essential if the Sun is rejuvenated through mass-loss Scenario 1. In principle, however, the required shielding could be achieved by placing a large solar sail (see Figure 2.15) between Earth and the Sun. In the rejuvenation process in which L/d^2 remains constant, Earth gradually moves outward in its orbit around the Sun. In this situation the flux of UV radiation reaching Earth is reduced by a factor proportional to d^2, resulting in a much more modest increase in the received UV flux. This smaller UV flux increase tends to favor the solar rejuvenation process that keeps L/d^2 = constant (mass-loss Scenario 2) over that which keeps $L(X)$ = constant (mass-loss Scenario 1).

The Extended Solar System Lifetime

The solar rejuvenation process not only combats the natural aging of the Sun, quashing its bloated red-giant stage, it also extends the Sun's lifetime. Eternal youth, however, cannot be gained and, eventually, the Sun will become a helium-rich white dwarf. How much additional lifetime the Solar System will gain through rejuvenating the Sun, however, will depend upon the exact amount of mass lost, the amount of chemical mixing, and the quantity of non-thermal pressure support that can be engineered.

The main-sequence lifetime of a fully mixed star can be compared to that of a similar mass star evolving in the canonical fashion, and it transpires[11] for Sun-like stars that in the zero mass-loss situation $T_H \sim T_{MS}$, where T_H is the hydrogen-burning lifetime of the fully mixed (homogeneous) star. Complete mixing alone, therefore, doesn't greatly extend the lifetime of the Solar System. This result comes about because a fully mixed star evolves at a higher luminosity than a non-mixed star. In this manner, while the mixed star has access to more fuel in the form of hydrogen, it consumes it more rapidly. Indeed, this is the reason why mass-loss must accompany the additional mixing being invoked in the solar rejuvenation process. The mass-loss causes the star to evolve at a lower luminosity (recall Figure 5.1), and it is this effect – combined with the greater fuel 'access' – that results in the Sun achieving a greatly enhanced hydrogen-burning lifetime.

If the mass-loss rate is assumed to be constant throughout the hydrogen-burning phase (as shown in Appendix A), then the ratio of the final mass to the initial mass of a fully mixed star is $M_f/M_0 = m(X = 0) = \exp[\,-N\,(\,Q/c^2\,)\,X_0\,]$, where $Q = 0.007\,c^2$ is the energy liberated per kilogram of stellar material by the PP chain of fusion reactions. The effect of mass-loss and homogenization on the hydrogen-burning phase of a Sun-like star[12] is shown in Table 5.5

Table 5.5. Hydrogen-burning lifetime improvement (T_H/T_{MS}) due to mass-loss and complete chemical mixing. The (assumed constant) mass-loss rate is parameterized according to N (first column), and the final to initial mass ratio is shown in the last column.

N	T_H/T_{MS}	M_f/M_0
25	1.1	0.88
50	1.2	0.78
100	1.4	0.61
150	1.6	0.48
200	2.0	0.38
250	2.6	0.29
300	3.5	0.23
400	7.5	0.14
450	12.2	0.11

Table 5.5 indicates that for the typical mass-loss values required to rejuvenate the Sun, its main-sequence lifetime might reasonably be extended by a factor of between 4 and 6. The increase in the main-sequence lifetime for a homogenized Sun-like star undergoing mass-loss is, with reference to Table 5.1, about the same as introducing a 15 to 20 percent non-thermal pressure support within its interior.

Mixing It Up

Energy is transported by bulk convective motion in the outer third of the Sun's radius, and in this region the material is well mixed and essentially homogeneous. The inner two-thirds of the Sun's radius, however, is not compositionally mixed at the present time. The problem that the future star engineer must solve, therefore, is clear: How might the inner regions of the Sun be chemically mixed? Huber Reeves,[13] in his book *Atoms of Silence*, captures the essence of the sought-after process by arguing that a pump of some type is required to cycle the hydrogen in the Sun's envelope to the central regions, where it can be consumed through fusion reactions. Even more prosaically Richard Cathcart[14] comments that the Sun must be stirred "much as one stirs a cup of coffee to mix the sugar and the liquid." Although the various analogies and the result required can be stated in a clear fashion, the actual physics of the problem are profound. There is simply no easy way to mix the interior of the Sun.

Hubert Reeves notes that additional mixing would take place within the Sun if a 'hot spot' could be created just above its fusion core. Indeed, such a process would work. However, the problem is, how might it be achieved? As shown in Chapter 3 the Sun's fusion core extends over the inner third of its radius (a region actually containing ~70 percent of the Sun's total mass), so the asteroengineer has to find a means of transporting energy deep into the Sun's interior. Directing asteroids or even large Kuiper Belt objects to impact upon the Sun will probably not produce the desired heating at the required depth. A numerical study carried out by David Andrews[15] of Armagh Observatory, Northern Ireland, indicated that a 14-km diameter asteroid (with a mass $\sim 5 \times 10^{15}$ kg) impacting the Sun's surface with a velocity of 600 km/s might penetrate to a depth of 20,000-km before becoming thermalized with the stellar surroundings. This penetration depth, which is of order $R_\odot/35$, is 10 times smaller than required. A larger asteroid will penetrate somewhat further, but not by an appreciably larger amount.

Ever resourceful and inventive, Reeves suggests that a high-powered laser might be used to generate a hot spot within the Sun, but again the problem is to get the energy to the right location deep within the interior. An alternative approach might be to develop methods to protect the intended impactor from being destroyed too soon. A combination of extremely high heat-resistant materials and strong enveloping magnetic fields might do the job, but the extent to which such futuristic technologies might be developed is hard to predict and currently unknown.

Black Hole Mixing

Black holes tend to be thought of as solely destructive entities and, indeed, they are at the heart of some of the most powerful and violent radiation-emitting sources that astronomers have ever detected. The million solar mass and larger black holes that power quasars and active galactic nuclei, however, are the leviathans of their race; the ones that the star engineer might be interested in are just a few hundredths of a millimeter in scale and weigh in at a relatively puny 10^{21} kg (about one one-hundredth the mass of our Moon). We know that millions of solar mass black holes do exist; it is also reasonably certain that stellar-mass black holes exist (such as in the X-ray binary system Cygnus X-1). But there is currently no clear observational data or consensus among astronomers concerning the existence of sub-stellar mass black holes. At this stage we shall assume that such black holes do exist, and that our distant descendants will work out ways of detecting, capturing, and transporting them through the solar neighborhood.

The question now is what happens to a low-mass black hole when carefully placed at the surface of a star? Simply put, it will begin to oscillate backwards and forwards across the star's interior.

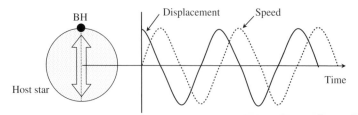

FIGURE 5.3. Displacement and velocity variation (scaled respectively to R_{star} and V_{max}) with time for a low-mass black hole placed at rest into the outer layer of a star. The solid line shows the variation of r/R_{star}, while the dotted line shows the value of $V(r)/V_{max}$. The period of oscillation is determined according to the distribution of mass within a star's interior.

It will move inward from its initial release point, gaining velocity as it moves toward the stellar center. It will pass through the center of the star (where it will attain its maximum velocity; see Figure 5.3), and thereafter it will begin to slow down as it journeys once again outward, just reaching the stellar surface at a location diametrically opposite to its release point. The whole oscillatory process will then repeat itself. Figure 5.3 illustrates the idea.

The motion of a black hole placed at the surface of a star is described (at least to a first approximation) by the same equation that accounts for the small oscillations of a simple pendulum. The variation in the displacement and velocity of such a simple harmonic oscillator (sho) are described by the equations: $r(t) = R_{star} \cos(\omega t)$, and $V(t) = V_{max} \sin(\omega t)$, where t is the time since release, $r(t)$ is the displacement from the center [with $r(0) = R_{star}$, where R_{star} is the radius of the star], $V(t)$ is the velocity [with $V(0) = 0$], V_{max} is the maximum velocity, and ω is the angular frequency.[16] Straightforward analysis reveals that the angular velocity is dependent upon the surface gravitational acceleration and the amplitude of motion. Accordingly, $\omega = (g/R_{star})^{1/2}$, where $g = G M_{star}/(R_{star})^2$ is the surface gravity of the host star (G is the universal gravitational constant). The period of oscillation T_{sho} is expressed according to this relationship: $T_{sho} = 2\pi/\omega$. For the Sun, $g = 273.76$ m/s^2 and $R_{star} = 6.96 \times 10^8$ m, which dictates that $T_{sho} \approx 10^4$ seconds (corresponding to about two and three-quarter hours).

Now, the result just presented is derived from the assumption that the Sun has a uniform and constant density throughout its interior; a more detailed calculation[16] allowing for the Sun's central mass concentration, however, reveals that the black hole oscillation time is, in fact, about 70 percent of T_{sho}, indicating an oscillation period of just under two hours duration.

A low-mass black hole moving through a star just once won't produce much of a mixing effect. In this engineering project, however, time is not considered to be of the essence, and the black hole will continue to oscillate back and forth across the star hundreds of millions of times, with each oscillation stirring the interior just a little bit more. Mixing will be initiated in the wake of the black hole once its speed of motion exceeds that of the local gas sound speed, since under such conditions a shockwave will be produced, and it is this effect that will produce the required interior mixing. An everyday analogy of the effect being envisioned (without the shockwave) is that of placing a small ball-bearing inside of a spray can of paint. By shaking the can the ball-bearing 'stirs' the paint and keeps it well mixed and fluid. A detailed set of calculations[16] indicates that a black hole placed within the Sun will be traveling at a maximum speed of something like 1,600 km/s when it reaches the Sun's center—a speed that is three times greater than the local sound speed. Indeed, the speed of a black hole oscillating within the Sun exceeds that of the local sound speed over 95 percent of its interior, only traveling at subsonic speeds in the outer 5 percent by radius.

A Steady Stellar Diet

With respect to reducing the mass of a star there are essentially two scenarios that might be applied. The mass can be either extracted from the surface, or it can be extracted from the deep interior. The first method essentially entails the inducement of an enhanced stellar wind, while the latter invokes accretion onto a central black hole.

A number of solar-mining and star-lifting scenarios have been described over the years, but here we will review just two of them. The first method has been described by Paul Birch[17] and involves a fleet of ramscoops. The second method was developed by David

Criswell[18] and involves the placement of multiple particle accelerators in orbit around the Sun.

The ramscoop method of solar-mining is illustrated in Figure 5.4. It was originally developed by Birch to extract just the heavy elements – or metals – from the Sun and, accordingly, the hydrogen and helium is not specifically collected. The ramscoop method could be extended, however, along the lines of the Bussard

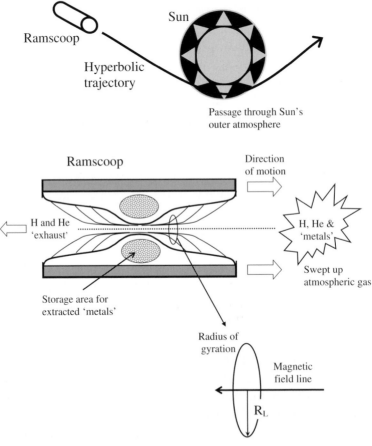

FIGURE 5.4. Solar-mining by the ramscoop method. The ramscoop is accelerated to skim through the Sun's outer atmosphere (upper figure). Ionized material entering the ramscoop's mouth will interact with a strong internal magnetic field and begin to move along spiral trajectories (center figure). The radius of the spiral trajectory is governed by the mass of the ion, and the more massive particles will have larger Lamor radii (R_L) (lower figure).

ramjet[19] to sweep up the hydrogen and helium in the Sun's outer atmosphere. The mining scenario begins with the ramscoop being accelerated to a fast (hyperbolic) orbit that allows it to skim through the Sun's outer atmosphere, thereby gathering material into its 'mouth.' The ionized material brought into the ramscoop's interior will interact with a strong, internally generated magnetic field, causing it to move along spiral trajectories. The radius of the spiral path is described by the Lamor or gyration radius, which increases in proportion to the mass of the trapped ion. In this manner the various ions can be sorted, one mass from another. By exploiting the ion-mass segregation process the various solar 'metals' can be siphoned off as they spiral along the magnetic field lines on their way to the ramscoop center.

Having passed through the Sun's outer atmosphere, the ramscoop returns to orbit with a diminished speed and an increased mass. Following a docking maneuver with an orbital storage ship, the ramscoop is unloaded and once again accelerated towards the Sun for another mass-extraction dive.

The mass-loss procedure described by David Criswell[18] is illustrated in Figure 5.5. In this mass-loss scenario a constellation of satellite accelerators are placed in a polar orbit around the Sun. Counter-directed and oppositely charged ion beams are then sent between each of the accelerators. In this fashion, the beams produce a loop of current around the Sun. By causing the plane of the accelerators (and the current loop) to rotate around the Sun's spin axis, two channels will be formed through which the Sun's outer atmospheric plasma will be continuously ejected. In this scenario the material will be predominantly swept outward in the plane of the Sun's equator.

An alternative scenario to that illustrated in Figure 5.5 has the accelerator satellites orbiting the Sun's equator, and in this configuration the current loop will cause material to be ejected from the Sun's poles. To induce mass-loss under this scenario, however, the polar regions need to be heated so the atmospheric atoms will have enough energy to escape the Sun's gravitational potential. Additionally, the orbital radius of the accelerator satellites could be made to systematically contract and expand, heating the Sun's atmosphere through the generation of sound waves. Criswell calls this mass-loss mechanism the "huff-n-puff" method.

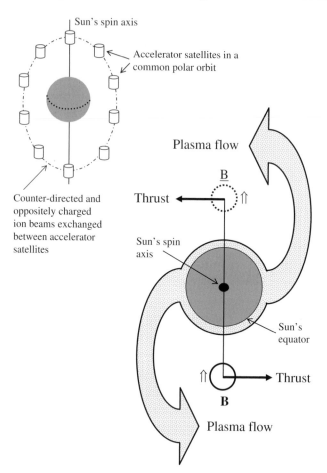

FIGURE 5.5. Equatorial mass ejection from the Sun. A constellation of accelerator satellites are placed into a common polar orbit around the Sun (upper left). By exchanging ion beams between each satellite a current loop BB is established (lower right). By rotating the plane of the accelerator orbit, surface solar material is driven outward along two 'exit' channels. (Diagram adapted from Criswell.)[18]

Martyn Fogg[20] has noted that, in principle, by varying the relative polar mass-loss rates a star might literally be lifted and propelled by a polar mass-ejection mechanism (Figure 5.6). Clearly the star-lifting scenario will have an important role to play in the re-direction of rogue stars (such as Gliese 710) away from a direct intercept path with the inner Solar System or our Solar System's Oort Cloud (see Figure 2.7).

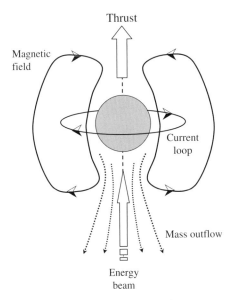

FIGURE 5.6. Lifting a star by the generation of an equatorial current loop. The star receives a thrust in the direction opposite to that of the material being ejected from the actively heated pole. (Diagram adapted from Fogg[20] and Criswell.)[18]

Brave New Worlds

To simply let the material extracted from the Sun dissipate into space would be a great waste of a useful resource. Although it is difficult to envision how the process might work, Criswell[18] has suggested that our remote ancestors might try to take the Sun apart layer by layer. If this approach can be made to work, low-mass dwarf stars could be constructed. As described in Chapter 3, the minimum amount of hydrogen that is required to produce a *bona fide* star is about 0.1 $M_\odot$. Following a solar rejuvenation scheme similar to the one outlined above, the future star engineer might, therefore, attempt to produce five to six dwarf stars as the Sun is modified. Such solar sibling stars would provide an immense wealth of additional energy for our descendants to utilize. By constructing Dyson sphere complexes around such sibling stars they could be turned into vast manufacturing centers or the parent stars to O'Neill-style space colonies.

The most likely region to assemble or eventually place (via star-lifting) solar dwarf star siblings will be in the inner Oort Cloud region. If a dwarf star of mass 0.1 $M_\odot$ is going to have a small to negligible effect on the orbits of objects out to, say, 1,000 AU from the Sun, then it must be placed in a solar orbit of radius ~1,500 AU. Such an orbit will clearly have a dramatic effect on the cometary nuclei in the inner Oort Cloud, but it is presumably safe to assume that by this advanced stage of engineering development, cometary impact avoidance will be a problem that has long been solved. A possible configuration of an orbit[21] for a full set of six solar sibling dwarf stars is shown in Figure 5.7 In the configuration shown each dwarf star leads and follows its nearest companions by 60 degrees. The separation between each dwarf star and its nearest neighbor is $\sqrt{3}\ R$, where R is the orbital radius around the Sun, and the maximum separation between any two dwarf star complexes is $2R$. Such a configuration minimizes the travel and communication times between the habitats in orbit about each of the dwarf stars, as well as with the planets in orbit around the Sun.

The main-sequence lifetime of a 0.1 $M_\odot$ star is measured in trillions of years, and (as illustrated in Figure 3.6) the fully mixed interiors of such stars naturally ensure that they won't undergo a red-giant phase (see Figure 3.11). Although we do not know how it might be achieved in practice, it seems that dwarf star production

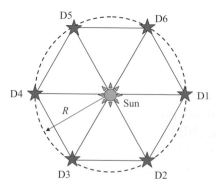

FIGURE 5.7. A possible arrangement of solar sibling dwarf stars. In this configuration the rejuvenated Sun has a mass of 0.4 $M_\odot$, and each dwarf star (D1 to D6) has a mass of 0.1 $M_\odot$ (Note, $M_\odot$ means the present mass of the Sun). Each dwarf orbits the Sun at a distance R and is separated from its nearest neighbors by a distance of ($\sqrt{3}\ R$).

and orbital husbandry should be a highly sought after end product of the solar rejuvenation process.

An Alien Beast Within

To some extent the notion of placing a black hole at the center of the Sun might sound like the scenario for the film *Alien* (20th Century Fox, 1979): "Lurking deep within the host star the black hole feeds from the inside outward, consuming the star inner layer by inner layer, ever hungry, until all has passed into its bloated maw." Well, this all sounds rather dramatic, but there are in fact good reasons to consider placing a black hole at the center of the Sun in order to rejuvenate it.

The possibility that the Sun might contain a low-mass central black hole was considered in some detail during the 1970s as a possible solution to the so-called solar neutrino problem.[22] The basic principle at work in this scenario is that the Sun can tap the energy liberated by material as it falls into the black hole. The more energy supplied by the accreting black hole, the less energy the Sun has to generate through fusion reactions and the lower its central temperature will be. The mathematical description of an accreting black hole placed within a star is given in Appendix B, and we refer the reader there for details. The main point for discussion at this stage relates to Equation (B.6), which describes star consumption time (literally, the time for the black hole to consume the star) $T_{consume}$ to the canonical main-sequence lifetime. If we once again assume that sub-stellar mass black holes with $M_{bh} \sim 10^{21}$ kg exist, and that such objects can be recruited in the Sun rejuvenation process, then $T_{consume}/T_{MS} \sim 0.1$ when the energy conversion efficiency (see Equation B.1) is 10 percent. This result tells us that in the case of the Sun, $T_{consume}$ is of order one billion years. Even if the black hole is inserted into the Sun near to the end of its canonical main-sequence lifetime it will not buy our distant descendants much additional time. If a highly optimistic energy conversion efficiency of, say, 25 percent is assumed, then the consumption time for the Sun is increased to about 2.5 billion years.

To increase the lifetime of the Sun well beyond its canonical main-sequence lifetime, it would seem that the only option for the star engineer is to develop techniques for controlling the accretion rate onto the black hole. The luminosity of a black hole radiating at its maximum rate is set by the Eddington limit,[23] which is related to the mass of the black hole: $(L_{Edd}/L_\odot) = 4.3 \times 10^4 \, (M_{bh}/M_\odot)$. From this relationship we see that once the black hole mass exceeds 2 $\times 10^{-5} \, M_\odot$, it will already supply enough energy to power the Sun at its current output level. A 10^{21} kg black hole will achieve a mass of $10^{-5} \, M_\odot$ some 400 million years after its insertion into the Sun (assuming 10 percent energy conversion efficiency). After this time the accretion rate would have to be carefully controlled and reduced so that $L_{acc} = L_\odot$, rather than $L_{acc} = L_{Edd}$. The Sun's central region is certainly an extreme environment from which to expect any Earth-produced machinery to work, but we are in the luxurious position of saying that this is a problem for our distant descendants to solve.

Although the controlled accretion onto a black hole at the center of the Sun can, in principle, quash the requirement for nuclear energy generation, the first problem to be solved is the placement of the black hole at the Sun's center. To a good initial approximation, however, the black hole placement can be modeled as a harmonic oscillator with an additional damping (or resistive) term. Under this approximation it can be shown[16] that the equation of motion has a straightforward solution with displacement $r(t)$ being described by the equation $r(t) = R_{star} \, e^{-Dt} \cos(\alpha t)$, where D is a constant and α is related via various constants to the angular frequency ω defined earlier for the simple harmonic oscillator. The constant D in the exponential term is a damping parameter that reduces the amplitude of oscillation. The larger the value of D the more rapidly the black hole will settle toward the center of the Sun. The timescale upon which the amplitude of motion decreases is given by the e-folding timescale defined as $T_{e-fold} = 1/D$. For each time interval corresponding to T_{e-fold} the amplitude variation decreases by a factor of $e \approx 2.718$.

If the damping of the black hole's motion is solely due to accretion, then it turns out that the e-folding time is of order many tens of millions of years. Such a settling timescale may be thought far too slow[24] by future star engineers and, consequently, steps

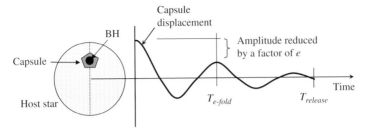

FIGURE 5.8. Delivering a low-mass black hole to the center of a star. This schematic diagram shows the black hole initially contained within a capsule designed to produce a large damping effect. To the right the displacement of the capsule with time is indicated. The amplitude of oscillation decreases by a factor of $e = 2.718...$ on intervals corresponding to $T_{e\text{-}fold}$. The black hole is released at time $T_{release}$ after which it begins to accrete material from the host star.

could be taken to make the damping term D as large as possible, thus reducing the e-folding time. One way in which this could be achieved is to initially constrain the black hole to reside within a delivery capsule that has been specifically designed to produce a significant viscous damping effect (Figure 5.8). High damping might be produced, for example, by the delivery capsule having a large surface area. An everyday analog to this scenario is that of a time-release medication capsule. In this fashion the capsule rapidly settles to the center of the Sun on a timescale of perhaps hours to days.[16] Once it has settled to the center the capsule releases the black hole in place, and the accretion of stellar material can begin.

Solar Wrap Control

A solar wrap might be thought of as a shrunken Dyson sphere. Rather than having a radius of order 1AU, a solar wrap, as the name suggests, would envelop the Sun at a distance of about 10 solar radii. In addition, and in contrast to a Dyson sphere, a solar wrap is designed to reflect most of the energy it receives from the Sun back toward its photosphere. The essential idea of the wrap is to cause the Sun to heat its own outer layers. This is a recipe in which the Sun essentially cooks itself. By heating its outer layers, the Sun will ultimately be forced to reduce the amount of energy it generates within its interior. Indeed, if none of the Sun's energy is allowed to escape into space (an extreme example that isn't very practical, perhaps), it will eventually evolve an isothermal interior with no energy generation taking place within its core. This perhaps seems an odd state of affairs, but the argument builds upon the results discussed in Chapter 3 and the summary shown in Figure 3.12. The point, of course, is that if the Sun isn't generating energy within its core, then it isn't using up its hydrogen fuel supply, and as long as the wrap stops energy leaking out into space it will have an extended, potentially indefinite, lifetime. The key concept here is that a star only generates energy within its interior because it loses energy into space at its surface. Plug the surface 'leak' and the requirement to be hot enough at the center to fuse hydrogen into helium goes away.

The solar wrap scenario works by exploiting the negative feedback mechanisms that operate within a star. Recall from Chapter 3 that the stability of a star is maintained by the existence of a temperature gradient; the temperature at each layer inside a star is just right to provide the correct gas pressure to support the weight of overlying layers. Now, if the energy radiated at the surface of a star is stopped from escaping into space by a totally reflecting cover (i.e., a solar wrap), a number of effects will come into play. First, the outer layers themselves will be heated and will consequently expand somewhat. Second, the heated region will gradually advance inward, forming a constant temperature (isothermal) zone. An outer isothermal zone comes about because there is no requirement to transport energy outward. When a region is isothermal there is no temperature gradient (it is the same temperature throughout), so the star will have to reorganize its internal structure so that the density gradient can maintain the stability condition. Recall that the ideal gas equation indicates that the pressure is determined by both the temperature and the density.

Eventually the isothermal region will extend all the way to the center, and the star will adjust its internal density at each interior layer so that it remains stable. The temperature of the isothermal star will be adjusted downward until it is just below the threshold for nuclear fusion reactions to occur. It is in this manner that the effect of a solar wrap becomes clear. By being surrounded by a fully reflective cover, the Sun will eventually evolve into a stable, isothermal object requiring no internal energy generation, this later condition coming about because there is no energy lost into space. The negative feedback mechanisms described in Chapter 3 will still apply to the isothermal Sun and, in principle, it will remain in a dormant state until some energy is allowed to escape through the wrap. At this point the Sun will undergo an internal readjustment so that the central temperature is raised by just the right amount to enable nuclear fusion reactions to replenish the energy that has been lost into space.

A series of detailed calculations[25] indicate that if, for example, the Sun can be transformed into an isothermal state, with a constant internal temperature of 10^6 K (i.e., below the temperature for efficient fusion reactions) and a central density somewhere within the range 100 to 1,000 kg/m^3, then it will expand by a factor of about 5 ½ times its present radius. In relation to a Dyson sphere, therefore, a solar wrap is not especially large, being in comparison some 40 times smaller in radius.

A fully reflective solar wrap is clearly not much use, of course; the Sun still has to provide energy to heat the planets within the Solar System. This suggests that an actively maintained (or leaky) wrap will have to be developed such that controlled amounts of energy are reflected back into the Sun's outer layers as needed. If such a Sun-altering scenario can be made to work, and if it is to extend the Sun's hydrogen-burning lifetime, then it has the added bonus that the Sun's mass is not changed and, consequently, there is no migration of planetary orbits. The dynamic solar wrap scenario won't necessarily stop the red-giant stage from occurring. But slowing down the rate at which the Sun converts hydrogen into helium will extend the time interval over which life can survive in the inner Solar System.

What the Future Holds

If humanity is to have a long-term future, then it must tame the Sun. There has been a long tradition of simply assuming that the aging of the Sun will destroy Earth and drive humanity (or – as is more likely the case – a very, very small and elite subset of humanity) into interstellar space to drift like a race of landless gypsies hoping to find a new planet to colonize. This may, of course, come to pass, but it seems that this would be a failed future. That is, it fails humanity collectively and deprives countless billions of people a potentially rich and prosperous life within a Solar System full of resources.

In this chapter we have attempted to outline a number of possible options for rejuvenating the Sun. Some of the scenarios are more fanciful than others, but the key point here is that there are conceivable options and possibilities. The Sun can be tamed, and it is not inevitable that it will become a red-giant. Clearly, we do not know what will come to pass in the future, and there may be many other possible means of rejuvenating the Sun; who knows what our descendants 1 million, 10 million, and 100 million years from now will be able to do. The future holds great promise for humanity—provided that humanity is prepared to realize it.

Notes and References

1. James Lovelock, *The Revenge of Gaia*, Penguin Books, London (2006). pp. 44–45. Lovelock further argues that in about 1 billion years' time the Sun's luminosity will have reached a level 9 percent greater than it is now, and that Gaia, the self-regulating 'living' world-system, will die. Indeed, Lovelock argues that in as little as 100 million years from now the Sun's energy output will have risen to a level that will force the biosphere into a new 'hot-state.'
2. From Equation (3.12) it can be seen that the luminosity at the beginning and the end of the main-sequence requires that the mass be reduced by a factor $[(\mu_{begin}/\mu_{final})^{7.5}/(1 + X_0)]^{1/5} = 0.286$, given that $\mu_{begin} = 0.613$ and $\mu_{final} = 1.613$.
3. See Beech, M., Sensitivity in the models of massive stars. *Astrophysics and Space Science*, **161**, 133–143 (1989); Beech, M., and Mitalas, R., The homogeneous evolution of massive stars. *Astronomy and Astrophysics*, **213**, 127–132 (1989).
4. In this derivation we have made use of the binomial expansion, which for small values of δ gives $(1 + \delta)^{-1} \approx (1 - \delta)$.
5. George Greenstein. Order of magnitude 'theory' of stellar structure, *American Journal of Physics*, **55** (9), 804–810 (1987).
6. Martin Beech. Aspects of an asteroengineering option. *Journal of the British Interplanetary Society*, **46**, 317–322 (1993).

7. For further details see M. Beech, A novel stellar model: 'a sacrifice before the lesser shrine of plausibility'. *Astrophysics and Space Science*, **168**, 253–261 (1990).

8. The helium main-sequence is nicely described in the book by R. Kippenhahn and A. Weigert, *Stellar Structure and Evolution*, Springer-Verlag, Berlin (1990), pp. 216–218. Detailed calculations reveal that a $1M_\odot$ chemically homogeneous helium-burning star is about 300 times more luminous than a $1M_\odot$ chemically homogeneous hydrogen-burning star. Helium-burning stars have smaller radii and higher central and surface temperatures than hydrogen-burning stars of the same mass. The very high central temperature appropriate to helium-burning via the triple-α reaction (see Note 15, Chapter 3) dictates that the energy is carried outward from the center by convection rather than radiation in homogeneous, solar-mass helium-burning stars.

9. This point is further discussed in the research paper by M. Beech and R. Mitalas, Effect of mass-loss and overshooting on the width of the main-sequence of massive stars. *Astrophysical Journal*, **352**, 291–299 (1990).

10. The Stefan-Boltzmann law is used to determine the surface temperature of the Sun. Accordingly, $Te^4 \sim L/R^2$, where L is the luminosity and R is the radius. Now, Equation (3.5) indicates that $R \sim M/T_c$, where M is the mass of the star and T_c is the central temperature. To evaluate the central temperature we have solved the integral: $L = \int \varepsilon dm$, where ε is the energy generation rate per unit mass of stellar material, as given by Equation (3.11). Accordingly, $L \sim X^2 T_c^{6+\alpha}/\mu^2$ Typically, for solar-mass stars, $\alpha \approx 4$. Working through the substitutions, the surface temperature of the star can be shown to vary as $Te(X)/Te(X_0) = 1^{(1+2\gamma)/4} m^{-1/2}(1+x)^{-2\gamma} X^{-\gamma}$, Where $\gamma = 1/(\alpha+6)$.

11. Following Reference 7, $T_H(x) \sim T_{MS} [1 + x^{17}] (1 + X_0)/(17 q)$, where $q \sim 0.1$ is the mass fraction of hydrogen available to a star evolving canonically. At hydrogen exhaustion, $x = 1/(1 + X_0)$, and $[1 + x^{17}] \sim 1$, leaving $T_H \approx T_{MS} (1 + X_0)/1.7$, and since $X_0 \approx 0.7$ so, $T_H \approx T_{MS}$.

12. Building upon Reference 7, the main-sequence lifetime of a homogeneous Sun-like star with mass-loss is $T_{HML} = T_{MS} [(1 + X_0)/q] \exp(4 k) \int x^{17}\exp(- 4 k x) dx$, where the integral is from x to 1, and k is given in Equation (A.7).

13. H. Reeves, *Atoms of Silence: An Exploration of Cosmic Evolution*. MIT Press, Cambridge, Massachusetts (1985).

14. From the essay by R. B. Cathcart, True asteroengineering: intrusive sun-stoking rejuvenation macroprojects. See http://www.daviddarling.info/Cathcart.html.

15. A. D. Andrews, Investigations of micro-flaring and secular and quasi-periodic variations in dMe flare stars. *Astronomy and Astrophysics*, **245**, 219–231 (1991).

16. In the detail calculation we have assumed that the Sun has an $n = 3$ polytropic structure [see Kippenhahn and Weigert's book, Reference 8], and the equation of motion for the black hole is integrated numerically. The polytropic approximation allows for the mass distribution to be centrally condensed with, in the $n = 3$ case, the central density being some 54 times greater than that of the constant density model of the same mass.

17. Paul Birch, Supramundane planets. *Journal of the British Interplanetary Society*, **44**, 169–182 (1991).

18. David Criswell, Solar system industrialization: implications for interstellar migration. In *Interstellar Migration and the Human Experience*. R. Finney and E. Jones (eds.), University of California Press, Berkeley (1985). pp 50–87.

19. An interstellar ramjet (or ramscoop) was described by physicist, Robert Bussard in his paper, Galactic matter and interstellar flight, *Astronautica Acta*, **6**, 179–194 (1960). The spacecraft described by Bussard uses a large scoop for channeling interstellar material into a central 'fusion' chamber, where energy is generated to propel the ship. The faster the spaceship goes, the more interstellar material it can sweep up and, consequently, the spacecraft could, in principle, reach relativistic speeds.

20. Martyn Fogg, Solar exchange as a means of ensuring the long-term habitability of Earth. *Speculations in Science and Technology*, **12** (2), 153–157 (1988). Viorel Badescu and Richard Cathcart, Stellar engines for Kardashev's Type II civilizations. *Journal of the British Interplanetary Society*, **53**, 297–306 (2000) have introduced the term Class A engines for machines (processes) that ultimately produce a thrust force from a star. Class B engines, on the other hand, use the radiation emitted by a star to produce mechanical power. A Class C engine is a combination of both A and B, and Badescu and Cathcart argue that such machines will provide a Kardashev Type II civilization with both power and interstellar transport (such as in the gargantuan, Solar System-shifting sail processes envisioned by Leonid Shkadov.

21. Aesthetics more than anything else underlies the configuration shown in Figure 5.8. The system, however, is stable provided that the mass of each sibling star is the same. Indeed, each sibling will move around the circular orbit (circumscribed about the hexagon) in pace and equally spaced from its neighbors. In essence, the 'hexagon' (with a sibling star at each node) rotates about the center (the Sun) as

if it were a ridged structure. It can be shown, in fact, that any regular polygonal arrangement of equal mass dwarfs will follow a stable circular orbit about the central Sun. Three sibling dwarfs of mass $0.2\ M_\odot$ situated at the 'corners' of an equilateral triangle centered on the Sun would also be a possible stable configuration. For a good review of such orbits, see Eugen Butikov, Regular Keplerian motions in classical many-body systems. *European Journal of Physics*, **21**, 1–18 (2000).

22. John Bachall [*Neutrino Astrophysics*, Cambridge University Press, Cambridge (1989)] nicely summarizes the 'non-standard' stellar evolution solutions offered to solve the solar neutrino problem. The solar neutrino problem has now been resolved in terms of 'new physics,' relating to the flavor-changing behavior of neutrinos.

23. The Eddington luminosity is derived on the basis that the pressure support for a star is that provided by radiation alone: $P = P_{rad} = (1/3)\ aT^4$, where a is the radiation constant and T is the temperature. Accordingly, $L_{Edd} = 4\ \pi\ G\ c\ M/\kappa$, where κ is the opacity of stellar material. At the high temperatures required for radiation pressure to provide the support for a star, the electron scattering opacity dominates, and the Eddington luminosity becomes $L_{Edd}/L_\odot = 4.3 \times 10^4\ (M/M_\odot)$.

24. In this scenario the black hole is not required to mix the star and, accordingly, the aim is to place it at the Sun's center as quickly as possible.

25. In these calculations we have numerically integrated the differential equations describing the pressure and mass variation within a star as a function of radius. [See, i.e., Kippenhahn and Weigert's book (Reference 8) for these equations.] Setting the temperature and central density as parameters to be chosen, the equations are integrated outward from the center to a radius R^* at which distance a total mass of $M^* = M_\odot$ is enclosed. This procedure determines the size of the isothermal Sun for the chosen temperature and central density. I have assumed in the calculations that the equation of state is that of a perfect gas. The stability of the isothermal Sun models has been tested against the gravo-thermal catastrophe condition described by Donald Lynden-Bell and Roger Wood [The gravo-thermal catastrophe in isothermal spheres and the onset of red-giant structure for stellar systems. *Monthly Notices of the Royal Astronomical Society*, **138**, 495–525 (1968)]. For the perfect solar wrap considered in these calculations, stability is assured provided the central density is less than $\sim 10^3$ kg/m^3 when $T = 10^6$ K. If a higher temperature of say 5×10^6 K can be engineered, then central densities between 10^4 to 10^5 kg/m^3 are allowed, and the equilibrium radius is reduced to of order $2\ R_\odot$.

6. Stars Transformed

In the preceding chapter we examined the ways in which an aging Sun might be rejuvenated and how the timescale over which life might be supported within the Solar System can be prolonged. In this chapter the focus will shift toward the broader question, which asks whether any galactic civilizations might have managed to alter their parent stars into a long-lived state.

Revisiting Carter

Physicist Brandon Carter has argued (recall Chapter 1) that the most likely time for an intelligent species to emerge is that corresponding to T_{MS}, the main-sequence lifetime of the planetary system's parent star. In this sense most intelligent species are likely to appear, if indeed they appear at all, at a time close to that in which the habitability zone is about to dramatically shift outward in response to the parent star becoming a red-giant. Stars with planets that have ages close to their main-sequence lifetime limits, therefore, are perhaps the most likely candidates for which the effects of rejuvenation might be evident. Indeed, in these systems the imperative to rejuvenate the parent star is at its highest for any late-emerging intelligent civilization.

An Exoplanet Review

As of this writing, astronomers have detected 252 extrasolar planets in orbit around 204 stars,[1] and the numbers keep growing every month! Age estimates have been made for a good number of the parent stars[1,2] to exoplanetary systems and the age, expressed as a fraction of the main-sequence lifetime, for 123 of these stars is shown in Figure 6.1. The number of stars in each bin is not

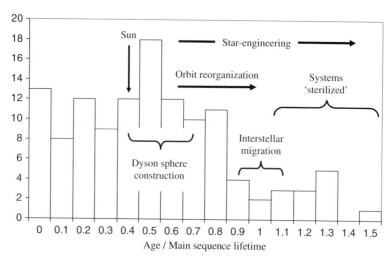

FIGURE 6.1. Age distribution of stars (expressed as a fraction of the main-sequence lifetime) known to support planetary systems. Superimposed on the diagram are the approximate time windows when planetary, as well as star-engineering and interstellar migration, might be initiated.

of specific relevance in this study, other than it being non-zero, but it can be seen that a complete main-sequence age spread has been sampled from $0.0 < \text{Age}/T_{MS} < 1.6$. About 56 percent of the planetary systems with published age estimates are associated with stars that are more than halfway through their main-sequence lifetime. There are 17 planetary systems in orbit around stars that have completed between 75 and 90 percent of their main-sequence lifetimes, and six systems that orbit stars that are within 1 percent of completing their main-sequence lifetime limit. The properties of the latter six systems are shown in Table 6.1.

Superimposed on Figure 6.1 are a set of suggested time windows when a civilization might wish to engage in the redesign of its planetary system. Clearly, we are many generations away from the initiation of any truly large-scale space engineering projects in our Solar System, and perhaps some 5 million years away from the initiation of any such endeavors. At this advanced time our Sun will be halfway through its main sequence. On this basis, the first possible appearance of engineering-related phenomena (i.e., Dyson spheres, or partial stellar wraps) in relation to planet-supporting systems might be seen when $\text{Age}/T_{MS} \sim 0.5$. As outlined in Chapter 4, if orbit manipulation or star-engineering

Table 6.1. Properties of the planetary systems with parent stars within 1 percent of their main-sequence lifetime limit. The last column refers to the study by Brian Jones and co-workers who looked at the possibility of stable orbits existing within each system's habitability zone (HZ). A 'Yes' indicates that stable orbits are possible; 'No' indicates that no stable planetary orbits are possible.

Star	Mass $(M_\odot)$	Age/ T_{ms}	$M_{Planet}/$ $M_{Jupiter}$	(a, e) of planet's orbit	Habitability
HD4308	0.83	0.98	0.05	(0.1, 0.0)	Yes
HD99109	0.93	0.98	0.5	(1.1, 0.09)	?
HD190360	0.96	1.07	1.5	(3.92, 0.36)	Yes
			0.06	(0.13, 0.01)	
HD213240	1.22	0.93	4.5	(2.03, 0.45)	No
HD216435	1.25	1.03	1.2	(2.6, 0.14)	No
70 Virginis	1.1	0.94	7.4	(0.48, 0.4)	Yes

options are going to be initiated then, in the case of our Solar System, these must start within the next 1 to 1.5 billion years (i.e., when the Sun has completed about 60 percent of its canonical main-sequence lifetime). If no large-scale engineering takes place within a planetary system that has reached the limit Age/T_{MS} ~1, interstellar migration might be initiated. And, again, if no system 'redesign' is initiated within a life-supporting planetary system, then once Age/T_{MS} > 1 for the parent star, the previously nurturing habitability zone will be sterilized and all life (if it evolved) will be killed off.

The fact that there are some stars with Age/T_{MS} > 1 in Figure 6.1 underscores the point made in Chapter 1 and Chapter 4, that parent star aging is a problem that must have already been faced by some galactic civilizations (although it must be emphasized that we do not know for certain if any of the systems listed in table 6.1 host planets capable of supporting advanced life forms). Brian Jones and co-workers[3] at the Open University in the UK have numerically investigated the possibility that Earth-like planets might exist in stable orbits within the habitability zones of most of the known planetary systems. In many cases they find that stable orbits can exist and, for the six systems listed in Table 6.1 (see last column), three might potentially support habitable planets. In addition, 26 of the 152 exoplanetary systems studied by Jones and co-workers with Age/T_{MS} > 0.65 allow for stable orbits to exist

Table 6.2. Exoplanetary systems that are known to have circumstellar dust disks. The disk extent data (fifth column) is taken from Trilling and co-workers.[4] The last column refers to the possibility of stable planetary orbits within the system habitability zone, as deduced by Jones and co-workers.[3]

Star	Mass ($M_\odot$)	Age/T_{ms}	N(planets)	Disk extent (AU)	Habitability
55 Cancri	1.03	0.6	4	~20 to 70	Yes
ρ CrB	0.95	0.6	1	~30 to 80	Yes
HD210277	0.99	0.7	1	~40 to 100	No

within their habitability zones. Time and improved ground-based observations will tell, of course, if there are any Earth-like planets within the habitability zones of the systems identified, but the possibility is certainly intriguing.

A survey of the properties of known exoplanetary systems having parent stars with Age/T_{MS} > 0.6 reveals that 55 Cancri, ρ Corona Borealis, and HD 210277 have associated circumstellar disks.[4] Table 6.2 presents a summary of these three systems. The presence of disks around these stars is certainly interesting for a number of reasons, not least from the point of view that they might be an indication of systems undergoing 'redesign' or star-engineering. It is not our intent to say that the exoplanetary systems with disks must support advanced civilizations in the process of large-scale engineering, but that the available data supports the possibility of them hosting ancient civilizations.

The fifth column in Table 6.2 indicates that the observed dust disks appear to begin at distances of between 20 to 40 AU from the parent stars. Within our Solar System the inner edge of the Kuiper Belt is located at about 40 to 45 AU from the Sun, so by analogy the exoplanets sporting dust disks apparently have Kuiper Belt-like regions composed of large ice/silicate objects of their own.

The Case of the Blue Stragglers

Sounding something like an adventure from the archives of Sherlock Holmes, the blue stragglers have long been a mystery to astronomers. They are oddities in that they have apparently evolved in a different way to their similar-aged and similar-mass companions.

The blue stragglers stand out from their neighbors as a consequence of the gregarious nature of stars. Rather than producing isolated stars – one at a time, here and there – nature shows a distinct preference for producing stars in relatively compact clusters. Indeed, dotted throughout the plane of our Milky Way galaxy, there are numerous open clusters[5] containing between a few hundred to several thousand stars. With characteristic dimensions of just a few parsecs across, galactic clusters provide astronomers with a very useful test bed for stellar evolution models. Since all the stars in a given cluster formed at the same time out of material with the same composition, and because they are all at essentially the same distance away from us,[6] then any observed variations in luminosity and temperature, from one star to the next, must be solely due to differences in their mass.

When a cluster Hertzsprung-Russell (HR) diagram is constructed, the lower mass stars will be located on the main-sequence, since the main-sequence lifetime of a star decreases with increasing stellar mass, while the more massive stars will have evolved to become red-giants. As the age of the cluster increases, so the mass of a star that can remain on the main-sequence (that is, not having evolved into a red-giant) decreases. Given the rapid evolution toward lower surface temperatures with the exhaustion of hydrogen within the core of a star, the cluster main-sequence develops a distinct turn-off point beyond which no stars will fall on the main-sequence. As the cluster ages, the luminosity and temperature of the turn-off point in the HR diagram decrease. The oddity of the blue stragglers is that they are cluster members found in the vicinity of the main-sequence, but beyond the turn-off point determined by their companion stars. Figure 6.2 shows the blue straggler region in a schematic cluster HR diagram.

With reference to Figure 6.2, the blue part of blue straggler refers to the fact that the stars are typically hotter than the main-sequence turn-off point, while the straggler part refers to the fact that other stars of the same mass, formed at the same time, have evolved into red-giants. Therefore, the question that follows is, why? What is it about blue stragglers that have made them evolve differently?

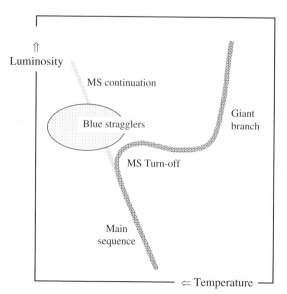

FIGURE 6.2. Schematic HR diagram for an old cluster of stars containing blue stragglers. The normally evolved stars delineate the main-sequence and the giant branch. The region where the blue stragglers are found is a continuation of the main-sequence, but beyond the main-sequence turn-off for the cluster. Some blue stragglers are also found to have higher temperatures than similar-mass stars located on the main-sequence. The observed location of blue stragglers in the HR diagram compares favorably with the predicted temperatures and luminosities expected for fully mixed stellar models as shown in Figure 5.2

Over the last 50 years a number of distinct formation scenarios have been developed for the blue stragglers. Among the suggested origins are the following:

1. They are stars that formed much later than their companions in the cluster.
2. They are young stars captured by the cluster.
3. They are stars that have accreted additional material from the surrounding interstellar medium, thereby increasing their mass, surface temperature, and luminosity, and have accordingly moved up the main-sequence.
4. They are coalesced stars produced by two stars actually colliding.
5. They are the result of a mass exchange between the two stars in a binary system.
6. They are stars that have undergone quasi-homogenous evolution.
7. They are stars containing a centrally accreting black hole.

Of the scenarios listed above the first three and last two seem to be the least popular among modern astronomers, at least as a general explanation, although there are no physical reasons to suppose that these scenarios could never apply. Scenarios 4 and 5 have been studied extensively in recent years and certainly appear capable of producing stars that qualify as blue stragglers. Scenarios 6 and 7 might, in some sense, be viewed as nature's version of the engineered stars that our descendants and advanced extraterrestrial civilizations might try to produce.[7] Craig Wheeler (University of Texas, Austin) introduced the term *quasi-homogeneous evolution*[8] to describe the partial extra mixing that was being invoked to produce the straggling effect. Full mixing of a star and red-giant phase elimination was not something that Wheeler was specifically trying to produce in his blue straggler models. In addition, as illustrated in Figure 5.1, a fully mixed star will eventually evolve 'blueward' to surface temperatures much hotter than that of the main-sequence. Most, but not all, blue stragglers have surface temperatures consistent with those expected for ordinary main-sequence stars.

There are a number of ways in which nature can apparently produce blue stragglers, with the cluster environment typically dictating which scenario operates most efficiently. Stars are more likely to collide and coalesce to produce blue stragglers in more densely populated clusters, for example, than sparsely populated ones. This being said, a star that has been engineered into a homogeneous, long-lived state (as described in Chapter 5) would also take on many of the characteristics of a blue straggler. In particular, a homogeneous star evolving with mass-loss such that the ratio L/d^2 remains constant (Figure 5.3) will evolve upward along the main-sequence, mimicking the blue straggler effect. In other words, an engineered star might, at least initially, be indistinguishable from a naturally produced blue straggler. Of course, not all blue stragglers have been engineered, but these stars are probably good candidates to study for additional signs that might betray an artificially engineered origin.

What characteristics might an engineered star have that would make it distinguishable from a naturally formed blue straggler? First, the environment in which the blue straggler is found could be a good indicator. It is highly unlikely, for example, that an isolated blue straggler, or one in a low-density cluster, will have been formed via a collision or through the capture and accretion of additional interstellar medium (in other words, Scenarios 1 to 4 listed above are unlikely to apply). Further, if no companion star is detected, then Scenario 5 can also be ruled out, leaving just the mixing and accreting black hole scenarios. The two mechanisms that are usually invoked to produce extra mixing are rapid rotation and strong internal magnetic fields. Both of these phenomena provide non-thermal pressure support within the interior of a star, and this enables the star to run at a slower evolutionary rate than a similar-mass star without additional pressure support. Certainly some blue stragglers show clear signs of supporting strong surface magnetic fields and some also show signs of rapid rotation, but the observations are by no means complete (or easy to obtain). Some blue stragglers also show clear signs of having undergone extensive, additional mixing.[9]

All in all, therefore, it is not entirely clear how an engineered star might look significantly different from a blue straggler produced through the natural mixing scenario, at least to begin with. Perhaps the best blue straggler candidates for further study with respect to the possibility of having been engineered are those with the highest surface temperatures (resulting from their being fully, rather than partially, mixed stars). If such stars also show indications of having strong magnetic fields (related to the magnetic mass-loss engines described earlier), and if they also show indications of mass-loss, then these stars might just be the ones undergoing rejuvenation by an advanced extraterrestrial race.

Recently, Orsola De Marco (American Museum of Natural History) and co-workers[10] have discovered four blue stragglers that have low-mass circumstellar disks with estimated radii of order 0.1 AU. Such disks might be the result of rapid rotation, or they might possibly be composed of the material being ejected from a star by a polar mass-loss engine (Figure 5.5). All the blue stragglers with disks are located within ancient globular clusters and are, therefore, already old stars. Should planets exist around such early-formed, relatively metal-poor stars, then their ancient heritage should favor the appearance of intelligent life.

The Time of Their Lives

In the last chapter it was argued that there was little advantage in seeding the Sun with a low-mass black hole in order to prolong its lifetime. Indeed, it is shown in Appendix B that the black hole mass increases exponentially with time, and the time for a low-mass black hole to consume the entire Sun is of order $T_{consume}$ ~1 billion years. For humanity this timescale is not of great use, but for an intelligent civilization emerging in a planetary system in orbit around a star more massive than the Sun, 1 billion years might be of great importance with respect to survival. Equation (B.6) reveals, for example, that $T_{consume} > T_{MS}$ once the mass of the parent star is greater than about 2 $M_\odot$. The black hole consumption time is nearly 2½ times greater than the canonical main-sequence lifetime for a 3 $M_\odot$ star. For civilizations that emerge in planetary systems with parent stars more massive than the Sun, the black hole seeding scenario might, therefore, buy a useful amount of time.

Martyn Fogg has developed a detailed model describing accretion-powered blue stragglers, and suggests that they might represent stars seeded with the primordial mini black holes postulated by Stephen Hawking (University of Cambridge, UK).[11] Although astronomers have not been able to confirm the presence of very low-mass black holes in interstellar space, some recent speculations concerning the formation of ball lightning on Earth may indicate that such objects really do exist. Although it is certainly a controversial proposal, plasma physicist Pace VanDevender has argued that long-lasting ball lightning events might be explained in terms of very low-mass, 1 (?) million kg,[12] black holes. VanDevender suggests that such minuscule black holes might form GEAs, the gravitational equivalent of an atom, in which the black hole acts as the nucleus around which atoms (rather than electrons) move within bound and stable orbits. All the above being said, the actual existence of low-mass black holes still remains highly uncertain at the present time, but should they exist – and provided a civilization can locate, transport, and then use one to prolong the main-sequence lifetime of its parent star – it is presumably among the more massive blue stragglers where such seeding might be evident. These are the stars with the shorter canonical main-sequence lifetimes and for which $T_{consume} > T_{MS}$.

This situation is probably not common, however, since the time for life to first appear and then evolve into an intelligent form around, say, a 2 $M_\odot$ star would have to be compressed into an uncomfortably short 1 billion years or less.

Under Construction

One of the problems inevitably encountered when trying to identify engineered stars is that many of the controlling processes that an asteroengineer might try to employ have natural counterparts. At some level, virtually all stars lose mass and have magnetic fields. Perhaps, therefore, the signs of ongoing stellar rejuvenation might be more readily identified by the debris and system reorganization being directed by the engineers. Among the various possibilities one can list are:

- Extensive Kuiper Belt-like and asteroid-belt-like debris clouds in older stellar systems – such debris resulting from asteroid mining and the deliberate rearrangement of terrestrial planet orbits.
- Bright optical flashes (periodic and non-periodic) due to near-specula reflection from large solar sails involved in the dynamical reorganization of system orbits (Figure 4.5).
- Periodic transits of a star by objects with distinctive, non-natural shapes (i.e., squares, louvers, and triangles), as suggested by Luc Arnold.
- Terrestrial planets with orbits that should be unstable given the age of the host star and the orbital radii of its companion planets. Such orbits would have to be continuously andactively maintained.
- Terrestrial planets with atmospheres betraying odd or even biotic chemistry, yet not situated in the habitability zone. Such observations would suggest terraforming in action.
- Rapid (milliseconds) brightness variations due to Dyson sphere islands transiting the disk of a parent star.
- Periodic X-ray emission from a Sun-like star due to the accretion of material onto a black hole undergoing oscillatory motion within its interior (Figure 5.3). The X-ray emission will be detected only when the black hole is close to the surface of the

host star, and the emission variation will have a period of about three hours.

- Stars with distinctive asymmetric mass-loss (Figure 5.6). Such an effect might be associated with a dwarf star close to or approaching a Sun-like star. The star-lifting of the dwarf would have been initiated to head off a close encounter with the Sun-like star's surrounding cometary cloud.
- Stars with unusually high abundances of helium relative to hydrogen. As a fully mixed star ages and the hydrogen is consumed throughout its interior, the He/H ratio will increase.
- Repeated brightness variations of a star associated with the passage of a giant ramscoop through its outer atmosphere (recall Figure 5.4).
- Stars with close-in (less than 1 AU) low-mass circumstellar disks produced by rapid spin induction or by a polar mass-loss engine (see Figure 5.5).
- Multiple dwarf stars occupying the same orbit (Figure 5.7) around a central star that is undergoing extensive mass-loss. The central star might also betray an unusually strong magnetic field due to the mass-loss driving engines.

There must be many more ways than those listed above by which the active and intelligent manipulation of a planetary system might be deduced. All we need to do is spend the time and effort to look, and then we need to look with a non-jaundiced eye. The Search for Extraterrestrial Intelligence (SETI) community has battled long and hard to achieve the levels of its present funding, and this is the situation in spite of SETI being one of the few scientific endeavors that has great public support. If we fail to look (and hope) for the signs of extraterrestrial civilizations within our galaxy, then we may fail ourselves and our imagination.

On the Threshold

Michel Mayor and Didier Queloz discovered the first exoplanet orbiting a Sun-like star in 1995. Surprisingly, however, the 0.47 Jupiter-mass planet was found to be orbiting its parent star – 51 Pegasus – at a distance of a mere 0.05 AU. Many such hot Jupiter-like planets have now been found, and these worlds are not likely to support intelligent life. To at least be consistent with what we know has occurred within our own Solar System, the goal of future studies is to find an Earth-like planet orbiting a Sun-like star at a distance of about 1 AU. There is no reason why such worlds shouldn't exist, but detecting them is far from easy.[13] We are, however, on the very threshold of being able to identify such planets around nearby stars.

In December 2006 the European Space Agency successfully launched its COROT satellite with the aim (at least in part) of detecting Earth-like planets in orbit around nearby stars. The mission name is derived from the title **CO**nvection **RO**tation and planet **T**ransits mission, and the satellite carries a 27-cm telescope specifically designed to detect planet-sized objects as they transit their parent star. When a transit occurs for a few short hours (depending on the orbit), it blocks a small fraction of its parent star's light from reaching the spacecraft detector. The satellite's instrumentation is able to record exceptionally small changes in the brightness of the star being monitored, and this should enable planets larger than about twice the size of Earth to be detected.[14]

The NASA *Kepler* mission is due for launch in late 2008 and, like COROT, it will search for sub-Earth-sized planets via transit observations. Named in honor of Johannes Kepler (1570–1630), the satellite will carry a 0.95-m diameter telescope and will be able to measure relative changes as small as 10^{-4} in brightness. During the spacecraft's four-year mission its telescope will stare at the same star field and continuously monitor 100,000 stars for the small, telltale brightness variations due to planetary transits.

The various satellite components that constitute NASA's Terrestrial Planet Finder (TPF) mission are currently scheduled for launch between 2014 and 2020. The suite of TPF satellites will be able to study all aspects of the formation and development of Earth-sized planets around nearby stars. Not only will the size, temperature, and location of any planets be determined, but their atmospheric chemistry will also be analyzed, thereby determining whether the planet might presently or someday support life. The detection of ozone or methane or both, for example, would indicate an active biotic system.

The mass range of terrestrial planets that might eventually be detected by future spacecraft mission will vary from perhaps a 10th

that of Earth (equivalent to a Mars-sized planet) to a maximum of about 10 times the mass of Earth.[15] Terrestrial planets that are smaller than Earth (i.e., Mars-sized planets) will have a low surface gravity and a more rigid crust, upon which high mountain ranges and deep valleys might form.[16] The large surface area to volume ratio for smaller planets will also result in their relatively rapidly cooling, leading to a decline in volcanic activity and active crustal recycling (i.e., as required by the carbon cycle; see Figure 2.24).[17] Such planetary characteristics are probably not conducive to the emergence and evolution of advanced life forms. Planets more massive than Earth, however, will have hotter interiors, higher surface gravity, and less rigid crusts. Under these circumstances the surface topology is likely to be muted, and maybe such planets will support deep (possibly global) oceans. These latter planetary characteristics are potentially more conducive to the appearance and evolution of advanced life forms. Intriguingly, William Dietrich and J. Taylor Perron (University of California, Berkeley) have recently argued that the existence of life might be betrayed on Earth-like planets by the kinds of topological features that they reveal.[18] Specifically, they point out chemical reactions precipitated by biotic activity will have potentially measurable, short-term effects on such processes as rock weathering, soil formation, upland slope stability, and river system dynamics.

As of this writing no nearby Sun-like star has been found to harbor a terrestrial planet, nor has any data been obtained to demonstrate the clear-cut existence of life having evolved anywhere in the universe other than on Earth. What is remarkable, however, is that we are literally on the threshold of possibly making such discoveries. Within the next 25 years humanity might actually know the answer to the age-old questions concerning the existence of other Earths and the evolution of ecosystems other than our own.

Notes and References

1. The most comprehensive website on exoplanetary discoveries and related publications is located at http://exoplanet.eu.

2. A set of age estimates for planet-supporting stars has recently been published by Carlos Saffe and co-workers [On the ages of exoplanet host stars, *Astronomy and Astrophysics*, **443**, 609–625, (2005)]. To convert the system ages into main-sequence lifetime fractions, recall from Chapter 2 [Equation (2.1)] that the main-sequence lifetime depends upon both the mass and luminosity of a star. Also recall, however, from Chapter 3 that main-sequence stars satisfy a mass-luminosity relationship. In this manner, the main-sequence lifetime can be written in terms of the mass alone. The observations, nicely summarized by Dr. R. C. Smith [An empirical stellar mass luminosity relationship. *The Observatory*, **103**, 29–31 (1983)], indicate that for stars more massive than 0.4 $_\odot$, the luminosity varies as: $L/L_\odot = (M/M_\odot)^4$. Accordingly, for Sun-like stars the main-sequence life time will vary according to the relationship: T_{MS} (yr) $= 10^{10} (M/M_\odot)^{-3}$.

3. Brian Jones and co-workers, Habitability of known exoplanetary systems based on measured stellar properties, *Astrophysical Journal*, **649**, 1010–1019, (2006).

4. D. E. Trilling et al., Circumstellar dust discs around stars with known planetary companions. *Astrophysical Journal*, **529**, 499–505 (2000).

5. In addition to the galactic (or open) clusters, the galaxy also hosts an extensive halo of globular clusters. The globular clusters are composed of the oldest known stars, with ages ~12 billion years, and they contain ~500,000 stars in a region that is ~20 pc across.

6. This condition holds provided that the size of the cluster D is small compared to its distance d from us. The closest galactic cluster to the Sun is that of the Hyades (visible to the naked eye in the constellation of Taurus), which is about 46 pc away. For this cluster, estimated to be about 800 million years old, D/d ~0.25.

7. These points were first raised in the article, Blue stragglers as indicators of extraterrestrial civilizations? *Earth, Moon and Planets*, **49**, 177–186 (1990).

8. The seminal theoretical paper on the blue straggler phenomena is that by J. Craig Wheeler, Blue stragglers as long-lived stars, *Astrophysical Journal*, **234**, 569–578 (1979). The idea of quasi-homogeneous mixing was further studied in the paper by Hideyuki Saio and Craig Wheeler, The evolution of mixed long-lived stars, *Astrophysical Journal*, **242**, 1176–1182 (1980).

9. A key indicator of extensive mixing is the lithium abundance of blue stragglers. The fusion reaction $^7Li + p ==> {}^4He + {}^4He$ + energy can proceed efficiently once the temperature is above just a few million degrees. The relatively low fusion temperature for ^7Li dictates that it is rapidly destroyed once it is mixed into the interior of a star.

In a number of observational studies it has been found that the lithium abundances of the blue stragglers within a cluster are significantly lower that that of the main-sequence stars near the turn-off point. These observations suggest that the blue stragglers have indeed undergone some form of additional mixing. A comprehensive review of the observations relating to blue stragglers is provided by L. L. Stryker in Blue stragglers, *Publications of the Astronomical Society of the Pacific*, **105**, 1081–1100 (1993).

10. See Orsola De Marco et al., A spectroscopic analysis of blue stragglers, horizontal branch stars, and turnoff stars in four globular clusters. *The Astrophysical Journal*, **632**, 894–919 (2005).

11. Martyn Fogg, Accretion-powered blue stragglers. *Speculations in Science and Technology*, **13** (1), 20–25 (1990). Hawking initially suggested that the solar neutrino problem might be solved by the Sun containing a primordial mini black hole [Gravitationally collapsed objects of very low mass, *Monthly Notices of the Royal Astronomical Society*, **152**, 75–78 (1971)], but this solution is no longer required since the problem has been solved in terms of a new physical understanding of neutrino characteristics. Hawking has further argued that low-mass primordial black holes must eventually evaporate, and produce what is known as Hawking radiation. The evaporation time for a 5×10^{12} kg black hole (which would have a size similar to that of a proton $\sim 5 \times 10^{-15}$ m) is about the current age of the universe.

12. VanDevender's ideas are discussed in Hazel Muir's article, Black holes in your backyard [*New Scientist Magazine*, 23/30 December (2006)]. With respect to the non-detection of Hawking radiation, VanDeveren argues that since there is currently no accepted theory of quantum gravity there is, likewise, no compelling reason to suppose that low-mass, primordial black holes must evaporate.

13. In a recent study published by Martyn Fogg and R. P. Nelson [On the formation of terrestrial planets in hot-Jupiter systems, *Astronomy and Astrophysics*, **461** (3), 1195–1208 (2007)] it is shown that material capable of forming terrestrial planets is able to survive within planet-forming disks, even after a Jupiter-mass planet (which formed in the outer part of the disk beyond the boundary where ice can form) has migrated inward towards the parent star. Fogg and Nelson conclude, "hot-Jupiter systems are likely to harbor water-abundant terrestrial planets in their habitability zone." This result is an important negation of the 'rare Earth' hypothesis that has been advocated by some scientists in recent years.

14. A number of planets have now been discovered through the transit detection technique (see Note 1). The first transiting planet was

found to orbit the star HD209458 [R. A. Wittenmyer et al., System parameters of the transiting extrasolar planet HD209458b. The *Astrophysical Journal*, **632**, 1157–1167 (2005)]. By studying the brightness variations over repeated transits it is now known that the radius of the 0.66 $M_{Jupiter}$ mass planet is 1.35 $R_{Jupiter}$. The planet is slightly inflated for its mass due to its close-in, 0.045 AU radius orbit around the parent star.

15. A study by Diana Valencia and co-workers [Internal structure of massive terrestrial planets, *Icarus*, **181**, 545–554 (2006)] finds that the size of a massive Earth-like planet scales as its mass M according to the relationship $M^{0.27}$. The radius increases, therefore, by about a factor of 2 when the mass increases by a factor of 10. If a planetary embryo grows to a mass greater than about 10 times that of Earth, it will begin to accrete and retain an extensive envelope of hydrogen and helium, resulting a gas giant (Jupiter-like) planet.

16. We know, for example, that Olympus Mons on Mars is nearly twice as high as Mount Everest on Earth. Likewise Valles Marineris is much larger and deeper than any similar such tectonically formed rift valley found on Earth.

17. The expected properties of various-sized Earths are described by J. J. Lissauer, How common are habitable planets? *Nature*, **404**, C11–C14 (1999). See also Note 11.

18. W. E. Dietrich and J. T. Perron. The search for a topological signature of life. *Nature*, **439**, 411–418 (2006). Climate change (as well as local weather variations) is also dependent upon biotic activity. A climate control example is that of the biotic sequestration of carbon to form carbonate rock deposits, as observed on Earth (see Figure 2.18).

7. Between Now and Then

Predictions concerning the future development of technology have, throughout recorded history, been notoriously poor, and this will probably always be the case. The future cannot be read, but it can be dictated to, and this is humanity's best hope for long-term survival. Indeed, our actions in the here and now will both influence and shape future events. The distant future of our descendants is not disconnected from our present.

The scale of engineering capabilities in the far future is entirely unknown to us. Certainly we can try to guess what might be, and without doubt we can try to steer research directions. But history tells us that it is more often the unexpected discovery that enables truly revolutionary advances. There is no obvious reason why such serendipitous advances should not continue to come about in the future, which begs the question, do we (that is, all humanity) actually have a long-term future?

Lord Sir Martin Rees,[1] former president of the Royal Society of London, and an authority certainly worth listening to, has placed the odds of humanity surviving into the 22nd century at no better than 50/50. Oxford University philosopher Nick Bostrom[2] concurs with the estimate of Rees and suggests that our chances of surviving this century are perhaps even lower, but probably better than 1 in 4. Bostrom further writes, "The most serious existential risks for humanity in the 21st century are of our own making. More specifically, they are related to anticipated technological developments." Thus, more specifically advanced technology will either be our salvation, or it will be our doom.

Do We Have a Near-Term Future?

The philosopher Jacques Derrida has argued[3] that the future "can only be anticipated in the form of an absolute danger.... [It] can only be proclaimed, presented, as a sort of monstrosity." Derrida's rather dire outlook may only relate to the near-term future. As of this writing, for example, the journal *Science* has just published the results of an extensive, four-year survey on global fish stocks.[4] The study concludes that commercial fisheries will collapse worldwide by 2050. Here is a near-term future monstrosity which we can only hope doesn't come to pass. With respect to the distant future, however, say hundreds or thousands of years beyond the lifetime of any human being alive today, the consensus of thought must surely be one of hope, even of a future utopia. Unless we (that is, all humanity) have some collective sense or belief in a better future for our children and their distant descendants' children, then surely there is no point in conserving anything. Indeed, why should we not simply continue to consume, waste, use, and abuse all that we possibly can with absolutely no regard for the consequences? If, on the other hand, we really do have a desire to secure humanity's long-term future, then our collective attitude toward what we are doing to the world, the climate, and to our fellow human beings must change now, and we must begin to think, regard, and plan for the long-term.

What Price the Future?

The evolutionary line that has led to our emergence has never been broken. Not once has the chain that links us to the very first primeval life forms been severed. Indeed, remarkably to think, not once in its entire 3-billion-year evolutionary history has life been fully exterminated.[5] There have always been survivors. And, as with the past, so with the future; there is no specific reason to think that the tenacity of life should weaken or wither as we move into the deep future. The indomitable Charles Darwin summed up the situation beautifully in *The Origin of Species*, published in 1859: "Of the species now living very few will transmit progeny of any kind to a far distant futurity.... As all living forms of life are the lineal descendants of those which lived long before the Silurian epoch, we may feel certain that the ordinary succession of generations has never been broken, and that no cataclysm has desolated the whole world. Hence we may look forward with some confidence to a secure future of equally inappreciable length."[6]

A billion years from now, even 4 billion years from now, life will still exist on Planet Earth. Whether any of that distant life will be directly derived from those human beings alive today, however, is very much an open question. For the first time in history, starting from about 50 years ago, humanity has had the power to destroy itself. Indeed, the weapons of mass destruction are now multitude (Figure 7.1).

The threat of an uncontrolled catastrophe was certainly on the minds of some of the scientists who pioneered the development of the atomic bomb. Before the first Trinity bomb was exploded

FIGURE 7.1. The atomic bomb explosion over Nagasaki, Japan, on August 9, 1945. As a result of this one detonation some 74,000 people were killed and a futher 75,000 people injured. By modern standards the Fat Man bomb that decimated Nagasaki was a mere firecraker.

near Alamogordo, New Mexico, on July 16, 1945, a series of approximate calculations were made by Hans Bethe (later a Nobel Prize-winning physicist) and Edward Teller (later an Ig Noble Peace Prize winner[7]) to show (or was it to reassure themselves and others?) that a nuclear detonation wouldn't ignite the Earth's atmosphere. Teller co-authored, in 1946, a more detailed report[8] for the Los Alamos administration in which it was surmised (after the fact) that "[N]o self-propagating chain of nuclear reactions is likely to be started" in the atmosphere directly heated by a nuclear blast. Indeed, the report comments, "The nuclear reactions most to be feared in air are the reactions of pairs of nitrogen nuclei." This is a remarkable comment and, although the report concludes that N + N reactions are highly unlikely to propagate within the atmosphere (and, indeed, this is the case), one wonders what the threshold for safety was considered to be. If the risk for ignition had been, say, as high as 0.01 percent, would the program of bomb development have been allowed to carry on?

In the modern era a number of new technologies are actively being pursued and developed, all of which pose potentially global disasters. Examples of such activities are the weaponization of space,[9] nanotechnology, and biological engineering.[10] Not that such technologies should be stopped, but are we sure they are being pursued safely and with due concern for their long-term consequences? Again, by 'safely' we are not referring to laboratory accidents or leaks, but literally the unintentional development of diseases and nano-robots that could destroy us. We tinker with such technologies at our own risk and, it appears, with absolutely no regard for the distant future consequences of such developments.

What are acceptable risks and who gets to decide what is acceptable are important questions, yet humanity seems to be incredibly under-prepared to deal with such issues. Important climate change and global warming issues, clearly demonstrated to be a real phenomena, are continuously downplayed by some governments and self-serving, industry-sponsored lobbyists. Clearly, for some, the devastation of Earth and the impoverishment of great swaths of humanity along with it is a worthwhile price to pay to achieve short-term profits or certain political goals. The inexcusable mindset that views the future as someone else's problem – if it continues – will destroy humanity, and we probably deserve no better. The time to think of the future is now. For example, Margaret Beckett, the British Foreign Secretary, nicely summed up the situation with respect to climate change when, during an address to the U.N. General Assembly on September 21, 2006, she said, "While it will not cost the Earth to solve climate change, it will cost the Earth, literally and financially, if we don't."

Thinking Long-Term

The rejuvenation of the Sun, the terraforming of Mars and Venus, the colonization of the Solar System at large are all set in the far future, but to realize that future humanity must begin to act and organize now. Again, Oxford philosopher Nick Bostrom has argued[11] that humanity should not only begin to think long-term but it should also maximize the pace of 'safe' technological development. Part of Bostrom's point is that any delay – even by just a few years – in the colonization of our Solar System (and perhaps beyond) has direct and negative consequences for the potential well-being of our distant descendants. Indeed, the future potential existence of innumerable sentient beings living worthwhile and fulfilled lives is denied every time we, in the here and now, accept short-term gains over long-term investments.

One of the most encouraging attempts at forward thinking and planning to emerge in recent years is that by the Long Now Foundation.[12] Its purpose is to replace the inherently sad and often morally bankrupt faster/cheaper economic policies that seem to permeate our current world with a philosophy of stewardship, reflection, and appreciation of both the distant past and the distant future. One of its symbolic projects has been the construction of the Long Now Clock (Figure 7.2). The purpose of the clock is to break away from the current everything-now mindset of our present lives. The Long Now Clock 'ticks' once per year. The aim is to allow the clock to evolve and be maintained for at least the next 10,000 years with each generation passing on the responsibility of stewardship to their immediate descendants.

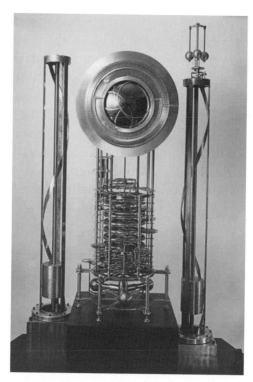

FIGURE 7.2. Prototype of the Long Now Clock. The clock is on display at the National Science Museum in London, England. The two outer columns contain the drive weights that power the clock, while the central column contains the binary mechanical computer (lower part) and the dial face (upper part). The dial shows the year as a five-digit number, as well as the sky locations of the Sun, Moon, and brighter stars. (Image courtesy of the Long Now Foundation)

Taking the Next Step

Astronomer Richard Gott III has argued that "the odds are against our colonizing the Galaxy and surviving to the far future, not because these things are intrinsically beyond our capabilities, but because living things do not live up to their maximum potential."[13] History, in other words, is against the human race surviving. Gott continues, however, "We should know that to succeed the way we would like, we will have to do something truly remarkable (such as colonizing space), something that most intelligent species do not do." Indeed, we would argue, Gott is absolutely right. For humanity to survive at all, it has to realize that there is a distant future and that the future begins now.

In Chapter 2 it was suggested that the first steps toward securing humanity's long-term future might be the development of a planetary defense initiative against direct impacts from near-Earth asteroids and comets. When might such an initiative begin? The scientific rationale for beginning an asteroid defense program has long been established, but the political and financial will has been found wanting. This, we would suggest, will possibly change in the not-too-distant future, but it might require the occurrence of a Tunguska-scale impact (recall Figure 2.2) before any serious political action is taken. Such an impact event will likely occur within the next few hundred years[14] (see also Figure 2.4).

What, then, are the possible time and engineering scales upon which the near-term future might be built? Table 7.1 represents a summary of possibilities.

Table 7.1. An outline scenario for humanity's colonization of the Solar System. The first column indicates the anticipated time interval, measured from 'now,' when various engineering options might be initiated. The second column indicates the engineering scale of the projects being undertaken. The third column offers some speculations on what might be achieved in each of the various time intervals.

Timescale (yrs)	Physical Scale (km)	Examples of Engineering Projects
Now	10^{-1}	International Space Station
$25 - 50$	10^0	First Moon base/solar sails/space tourism
$50 - 10^2$	10^1	First asteroid/comet defense systems deployed/main-belt asteroid mining
$10^2 - 10^3$	$10^1 - 10^2$	Terraforming of Earth's atmosphere/space elevator construction/asteroid colonies established/Moon settlements
$10^3 - 10^4$	$10^2 - 10^3$	O'Neill colonies/terraforming of Mars/terraforming of Venus/mining of Mercury
$10^5 - 10^6$	$10^4 - 10^5$	Greater Solar System colonization and the expanded utilization of its resources
$10^7 - 10^9$	$10^5 - 10^6$	Solar rejuvenation/stellar husbandry/planetary/orbit manipulation

In a remarkable series of studies[15] industry consultant Theodore Modis demonstrated that during the past two centuries humanity has ridden a remarkably stable 56-year repeating energy cycle. Cycle troughs correspond to times of great innovation and upheaval, while cycle peaks correspond to times of prosperity and economic boom. The peaks are inevitably followed by economic (especially so, given its present short-sighted nature) bust. Bruce Cordell has further argued[16] that, with respect to space exploration, the last Modis cycle peaked in the late 1960s – a time which saw the development of the *Apollo* Moon-landing program. If the Modis cycle holds true, the next epochs at which new exploration thrusts and innovations in space engineering are likely to be seen are in the decades[17] that contain the years 2025 and 2081.

At the risk of making predictions (even for the near-term future), we would suggest that perhaps during the 2025 Modis peak humanity will see the initial deployment of a space-based asteroid/comet impact avoidance network. In addition, the first use of large solar sails and the initial construction of what will become the first permanently occupied structures on the Moon will take place in this timeframe—the latter having already been announced as a NASA goal. Space-based tourism (currently an industry in the making) will also become an established public-sector business venture over the next quarter century. During the 2081 Modis cycle peak one might hope to see the development of a terrestrial space elevator,[18] the establishment of mining factories on the Moon and in the main-belt asteroid region, and the first human exploration of Mars. It is also likely that within this timeframe the first active steps toward the large-scale 'geoengineering' of Earth's atmosphere will take place. Those inspired to make wagers on such possibilities are invited to visit the Long Bets Foundation website at http://www.longbets.org/about. Beyond the end of this century, all bets are off on exactly when subsequent large-scale space engineering projects might be initiated. Indeed, the planning and construction timescales for these deep-future projects become very long, and certainly longer than that of a human lifespan. It does not seem unreasonable, however, to think that the terraforming of Mars might be completed during the projected 10,000-year time span of the Long Now project.

On November 20, 2006, Stephen Hawking was awarded the United Kingdom's highest academic honor – the Copley Medal of the Royal Society of London. In response to this event, Hawking commented upon humanity's future and argued that the human race, in order to secure its long-term viability, must move to a new planet beyond the Solar System. Hawking's sentiments are certainly sound, but humanity has absolutely no need to cast itself adrift in interstellar space to secure its long-term future. The extensive colonization of the Solar System and the active rejuvenation of the Sun make much more sense than simply abandoning home and heading for the stars. Interstellar space travel is full of uncertainty and provides no guarantee of a successful outcome. Not only this, as was suggested earlier, but the utilization and colonization of the Solar System will potentially benefit all of humanity, not just a select few.

We have deliberately avoided any discussion of the economic costs relating to the various engineering projects outlined in this book. Economics, at least in its present form, is entirely focused on short-term goals and to a certain extent we would also question, what price the survival of humanity? If we had the technological expertise and ability to avert a natural disaster that might otherwise result in the deaths of many millions of people, would that technology be held in check simply because it cost too much? We certainly hope not.

Future Earth

How many human beings can the world support? Currently of order 6 billion people eke out an everyday existence on every single continent of Earth (including the researchers who live and work in Antarctica). Humanity seemingly thrives and goes blindly about its business with little regard for the future. This has been our history. Humanity's 'footprint' is no longer irrelevant to the workings of nature, however, and we continue to destroy the environment at our own peril. Many estimates have been made concerning the number of human beings that Earth can reasonably support, with the numbers typically falling between 8 and 12 billion people.[19] Irrespective of the actual number (at this stage), projections on human population growth by the United Nations indicate that Earth's human-carrying capacity will be realized within the next 100 years (and probably within the next 50). This is another reason that humanity must begin to think of the future now; we can no longer assume that a 'business as usual' approach will work. The long-term future leading to the rejuvenation of the Sun certainly begins now, but the realization of that distant future is critically dependent upon what happens on this Earth within the next century.

If humanity manages to realize a distant future it is certain that those of us alive today will have virtually nothing in common with that future, other than lineage. Indeed, the people that might rejuvenate the Sun will likely know absolutely nothing about our world. Continents will have collided, split apart, and regrouped, and new mountain ranges will have formed and been weathered away many times over by the time that solar rejuvenation might begin (starting, perhaps, a few hundred million years from now).[20] The only common link between the future stewards of Earth and us will be an existence tied to the Solar System and a desire to secure the viability of the Solar System as a safe abode for life for as long as is humanly possible.

> We shall not cease from exploration
> And the end of all our exploring
> Will be to arrive where we started
> And to know the place for the first time
>
> T. S. Elliot, *Little Gidding*

Notes and References

1. Sir Martin Rees, *Our Final Hour*. Basic Books, New York (2003).
2. N. Bostrom, Dinosaurs, dodos, humans. *Global Agenda*, February, 230–231 (2006). See also: http://www.nickbostrom. com/.
3. J. Derrida, *Of Grammatology*. John Hopkins University Press, Baltimore (1967).
4. Boris Worm and co-workers, Impacts of biodiversity loss on ocean ecosystem services. *Science*, **314**, 787–790 (2006).
5. From the safety of much of the Western world this seems a rather trite statement. If one doesn't live in an Earthquake zone, an active volcanic region or on a seasonally hurricane-lashed coast, it is easy

to forget that nature has immense destructive power. An undersea landslip of just a few meters off the coast of Indonesia on December 26, 2004 resulted in the death of some 230,000 people along the coastal regions of the Indian Ocean basin. Indeed, 168,000 deaths occurred in Indonesia alone. This number is about the same as the entire population of the city of Regina, Canada. The staggering destruction that can be unleashed by nature is described in chilling detail by Bill McGuire in his book, *Global Catastrophes: A Very Short Introduction*. Oxford University Press, Oxford (2002).

6. From Charles Darwin, *The Origin of Species*. John Murray, Albemarle Street, London. Chapter 14 (1859). Darwin's complete text can be found at http://www.kellscraft.com/ index.html.

7. Teller's Ig Noble Peace Prize was awarded in 1991 in recognition of "his lifelong efforts to change the meaning of peace as we know it." More details can be found at: http://en.wikipedia.org/wiki/List_of_Ig_Nobel_Prize_winners.

8. The report by Teller and co-workers, Ignition of the atmosphere with nuclear bombs, *Los Alamos Technical Report LA602* is available at http://www.fas.org/sgp/othergov/ doe/lanl/docs1/00329010.pdf. Specifically it was the $N + N$ reaction that was considered in the report since nitrogen is the most abundant element within Earth's atmosphere.

9. The U.S. government, for example, has recently adopted an aggressive and strongly negative stance with respect to the banning of space weapons. In an unclassified strategic document [see, http://news.bbc.co.uk/2/shared/bsp/hi/pdfs/18_10_06_usspace.pdf], it is stated that "The United States will preserve its rights, capabilities, and freedom of action in space... and deny, if necessary, adversaries the use of space capabilities hostile to U.S. national interests." Mike Moore [Watch out for space command, *Bulletin of the Atomic Scientist*, **57**, 24-25 (2001)], discusses the apparent U.S. planning and build-up towards an arms race in space.

10. Lord Rees (Note 1) has highlighted the dangers posed by nano-technology and the development of 'superbugs' through biological engineering.

11. N. Bostrom, Astronomical waste: The opportunity cost of delayed technological development. *Utilitas*, **15**, 308–314 (2003). See also note 2.

12. The foundation's web page can be accessed at http://www.longnow.org/. A few of the thoughts presented in this section are discussed in M. Beech, The clock of the Long Now—a reflection. *Journal of the Royal Astronomical of Canada*, **101**, 4–5 (2007).

13. J. R. Gott III, Implications of the Copernican principle for our future prospects. *Nature*, **363**, 315–319 (1993).

14. Peter Brown and co-workers, The flux of small near-earth objects colliding with Earth. *Nature*, **420**, 294–296 (2002). The authors conclude that Earth is hit by a 10-Mton explosive equivalent energy object, which is something like the energy released by the Tunguska impact, at intervals of about 500 to 1,000 years.

15. Theodore Modis, *Prediction – Societies Telltale Signature Reveals the Past and Forecasts the Future.* Simon and Schuster, New York (1992).

16. Bruce Cordell, Forecasting the next major thrust into space. *Space Policy*, **12** (1), 45–57 (1996).

17. Modis has recently argued [The limits of complexity and chance. *The Futurist*, May-June (2003)] that the next major epochs of milestone innovation and growth will be circa 2038, 2083, and 2152. These epochs are not specifically related to space exploration (as studied by Cordell—see Note 16), but involve the appearance of new milestone technologies such as computers, the Internet, the automobile, transistor and integrated circuits, DNA sequencing, and so on, as witnessed during the past century.

18. A space elevator provides a physical link between Earth's surface and space. Stretching to some 35,786 km above Earth's surface, the geostationary point, the space elevator remains fixed above a specific location on Earth's equator. For further details, see: http://en.wikipedia.org/wiki/Space_elevator.

19. Joel E. Cohen, Population growth and Earth's human carrying capacity. *Science*, **269**, 341–346 (1995).

20. An issue not specifically addressed in this book relates to the gradual slowing down of Earth's geological activity. Indeed, as Earth's interior continues to cool off in the deep future, so surface tectonic activity will halt, bringing to a close, for example, the CO_2 cycle (see Figure 2.18). Earth's atmosphere will eventually need continuous manipulation and rejuvenation. This, however, will be an 'old technology,' honed through the terraforming of Mars and Venus, by the time the process is required to start on Earth. Likewise, the Moon will eventually slip out of a co-rotating state, as it drifts further away from Earth - the current rate of increase in its orbital semi-major axis being 38 mm / yr. Some commentators have seen this drift as a major issue, but it could be a trivial problem to solve by the time that action is actually required. The point being that many tens of millions of years from now, if humanity is still thriving, resetting the Moon's orbit through controlled flybys of asteroids, cometary nuclei, or Kuiper Belt objects will be a commonplace engineering process.

Epilogue

(With apologies to H. G. Wells)

Los Alamos Laboratories, New Mexico, 2145. It is lunchtime. Our gaze is directed towards a quiet corner of the otherwise busy refectory hall. "The Time Traveler" (for so it will be convenient to speak of him) was expounding a recondite matter to us.[1]

"I have seen the desolation that will be wrought upon Earth by the aging Sun." The response at the table was immediate. A multitude of crop-haired heads turned in unison. Conversations stopped in mid-sentence. The Time Traveler had our attention.

"As you all no doubt know," he continued, "The Osgiliath[2] Project unlocked the fundamental secrets of time travel two years back. But just four days ago, by Earth time, I took the very first human journey down a deep-future timeline."

We sat in stunned silence. This was news indeed. Certainly we all knew of the Osgiliath Project, and two Nobel Prizes had come to the laboratory as a result of it. But to actually send a human into the future—that was incredible. The official line was that only inanimate matter could be sent through time, and then only atomic-sized objects at that. A number of nanobots[3] had been sent through the time portal to verify that the process actually worked, but to think that the future had been revealed to human eyes, that was truly food for thought.

"Come on, that's impossible," one of the younger researchers taunted. "We all know that only non-sentient objects can be sent through the time funnel.[4] You're joking – right?"

"I traveled into the future. It is the truth," the Time Traveler replied, making no further effort to justify his fantastic announcement. His haunted expression, however, left us all a little uneasy. Indeed, upon reflection, it was clear that the trauma of seeing the deep future, and what lay in store for Earth, sat raw and indelibly stamped upon his face. "I initially traveled 4 billion years into the future," he eventually continued after a long, reflective silence. "I cannot convey the sense of abominable desolation that hung over the world. The red eastern sky, the northward blackness, the salt Dead Sea ... the uniform poisonous-looking green of the lichenous planets, the thin air that hurts one's lungs; all contributed to an appalling effect."[1]

Not one of us at the table moved. It was as if time stood still. We could hardly breathe. All eyes were directed towards the Time Traveler. Could his story really be true? Certainly, we all appreciated from basic astrophysics that the Sun must eventually go through a giant phase, but the consequences of that evolution were something that had never truly registered in any of our thoughts.

"I pushed deeper down the timeline," he eventually offered. "The Sun grew ever brighter; Earth became ever hotter." Then, with a look of anguish showing across his brow, he added, "The huge red-hot dome of the Sun had come to obscure nearly a tenth part of the darkling heavens. I looked about me to see if any traces of animal life remained, but I saw nothing moving in Earth or sky or sea. All the sounds of man, the bleating of sheep, the cries of birds, the hum of insects, the stir that makes the background of our lives—all was over." [1]

And so the narrative of the Time Traveler continued. We, his audience, sat dumbstruck, listening to his every word and revelation about the distant future and the eventual destruction of Earth, our home, by a bloated, red giant Sun.

"But is all this inevitable?" someone eventually asked. "Must this Sun-driven ruin be Earth's inevitable future?"

"No, it is not inevitable," the Time Traveler responded. "What I have seen need not come to pass. It is just the present future that I have witnessed." With these words the Time Traveler appeared to be done with his story. We all began to breathe again, none of us quite realizing how deeply absorbed we had become involved in the details being recounted. Bodies shuffled and stretched. A few people left the table, shaking their heads as they walked away; others sat in contemplative silence.

A few days after his lunchtime revelations, the Time Traveler disappeared. We never saw him again. Questions were asked, of course, and the police even investigated, but absolutely no trace of his whereabouts could be found. The Osgiliath Project was also closed down a few weeks after the Time Traveler disappeared, the accountants apparently finding the project to be too expensive, and of affording too few prospects for practical development and near-term investment expenditure recovery. Rumors abounded, as they always do in a place like this, that the project had gone underground and that the military boys were running the research now. Who knows? But one thing is for certain. After hearing what the Time Traveler had to say, a few of us that were seated around that lunch-time table have started to investigate ways in which the long-term husbandry of our Solar System might be achieved through the rejuvenation and engineering of the life-giving Sun.

Notes and References

1. Extracted from H. G. Wells, *The Time Machine*. Random House, New York edition (1931).
2. A name shamelessly taken from J. R. R. Tolkien. Osgiliath is Sindarin Elvish for 'citadel of the stars.' It was at Osgiliath that the chief palantir was kept, before being lost during the great civil war of Gondor. A palantir was a crystal globe that could show events from far away in both space and time. The history of Osgiliath is given in Tokein's *The Silmarillian* [Allan and Unwin, London (1977)].
3. Nanotechnology or technology on the nanometer (10^{-9}-m) scale will presumably be well developed by 2145, the imagined time at which the *Epilog* is set. Indeed, even today scientists at Columbia University have announced the development of a molecular spider that uses four, 10-nanometer long DNA 'legs' (the researchers conveniently appear to have forgotten that spiders actually have eight legs, but no matter) to clear a sterile path along a substrate. The specific research paper Behavior of polycatalytic assemblies in a substrate-displaying matrix is by Renjun Pei and co-workers, is published in the *Journal of the American Chemical Society* , **128** (39), 12693-12699 (2006). Neural computer networks incorporating biological 'brains' composed of cultured rat neurons have also been described recently by Thomas DeMarse and co-workers in, The Neurally controlled Animat: Biological brains acting with simulated bodies. *Autonomous Robots*, **11**, 305–310, (2001).
4. I have no real idea what to call the time 'funnel,' but I have always liked Kurt Vonnegut's expression "chrono-synclasic infundibulum," as used in his 1972 play *Between Time and Timbuktu*, or *Prometheus Five*. An infundibulum (of course) is something that is funnel-shaped.

Glossary of Terms

Astronomical Unit (AU) The semi-major axis of Earth's orbit around the Sun, corresponding to a distance of about 150 million km.

Blackbody radiator An idealized object that radiates electromagnetic energy into space in accordance with Wien's law and the Stefan-Boltzmann law.

Black dwarf A near fully cooled-off white dwarf.

Brown dwarf An object with a mass of between ~0.01 and ~0.1 $M_\odot$. Such objects are not massive enough to initiate hydrogen fusion reactions within their interiors.

Chandrsekhar limit The limiting mass for a stable white dwarf star – $M_{WD} \leq 1.4\ M_\odot$. The limit is set according to the degenerate electrons acquiring relativistic speeds.

CNO cycle The catalytic fusion reaction that enables the conversion of four protons into a helium nucleus with the liberation of energy. The process begins with the reaction $^{12}C + P \Rightarrow {}^{13}C$.

Croll-Milankovitch cycle Climate changes that are driven in response to small variations in the size, eccentricity, and orientation of Earth's orbit.

Degenerate matter Matter that behaves according to the Pauli exclusion principle.

e-**folding time** The time required to reduce a quantity by a factor of $e = 2.71828....$

Gaia hypothesis An idea developed by James Lovelock which views Earth's biosphere as a self-regulating system that operates via a complex series of interactions between its physical, chemical, and biotic components.

GRB Gamma-ray burst. This event is believed to be associated with a hypernova, which is characterized by the release of large quantities of very-short wavelength electromagnetic radiation (i.e., γ-rays and X-rays).

Greenhouse effect The process by which a planetary atmosphere is heated by the absorption of infrared radiation emitted by the planetary surface (after it has been heated by sunlight).

Hydrostatic equilibrium The achievement of a dynamical balance between the gravitational force, which is trying to collapse a star, and the outward (hot gas) pressure forces that are trying to expand a star.

Hypernova The final disruption stage of a massive, rapidly rotating star.

Kuiper Belt A disk-like distribution of primarily icy objects orbiting the Sun out to distances of several thousands of astronomical units. The existence of these objects was predicted by Gerald Kuiper in 1951. Pluto is the closest Kuiper Belt object to the Sun, while Eris is currently the largest known such object.

Lamor radius The radius of gyration for a charged particle moving along a helical path along a magnetic field line. For a given magnetic field strength the gyration radius increases in proportion to the particle mass.

Malthusian catastrophe The collapse of a population caused by it outgrowing the capacity to feed itself.

Metals A collective term applied by astronomers to those elements other than hydrogen and helium.

NEA Near-Earth asteroids. Those asteroids with orbits bringing them periodically close to Earth.

NEOs Near-Earth objects. The collective name given to those asteroids and comets that periodically pass close to Earth.

Neutron star The remnant of a star initially more massive than eight times that of the Sun that has undergone supernova disruption. Such stars are supported against gravitational collapse by degenerate neutron pressure.

Oort Cloud An extensive cloud of many trillion (or more) cometary nuclei that surrounds the Sun and delineates the outer boundary of the Solar System at distances of more than 100,000 astronomical units. The existence of this cometary cloud was first suggested by Jan Oort in 1950.

Parsec (pc) The distance at which an object would have a stellar parallax of one arc second.

Pauli exclusion principle (PEP) A quantum mechanical effect, first described by Wolfgang Pauli in 1925, which forbids two or more electrons from occupying the same phase space. A gas obeying the exclusion principle is said to be degenerate.

Photosphere The outermost region of the Sun where photons can escape into space without further atomic interactions.

Planetary nebula An ionization nebula produced by a star that has exhausted helium within its core and is evolving into a white dwarf.

PP chain The fusion reaction that enables the conversion of four protons into a helium nucleus with the liberation of energy. The process begins with the reaction $P + P \Rightarrow D + energy$.

Quarks Fundamental particles that carry a fractional electrical charge. Quarks come in six 'flavors,' and they are either found in groups of two (making up so-called mesons) or three (making up so-called hadrons). The proton is composed, for example, of two 'up' quarks and one 'down' quark.

Spectral classification A classification scheme based upon measured spectral lines and which characterizes stars according to their surface temperature.

Stefan-Boltzmann law One of the fundamental laws pertaining to blackbody radiators. The law dictates that the energy radiated into space per second per meter squared (i.e., the flux of electromagnetic radiation) varies as the temperature of the blackbody radiator raised to the fourth power.

Stellar parallax The apparent motion in the position of a star, over a six-month interval, due to Earth's motion around the Sun

Strange stars Hypothetical stars made of strange quarks held together by gravity rather than the strong interaction force of elementary particle physics.

Terraforming The process by which a planet is made habitable.

Torrino scale A scale developed to describe the impact threat associated with near-Earth asteroids and comets.

Triple-α reaction The fusion reaction that enables the conversion of three helium nuclei into a carbon nucleus, liberating energy in the process.

White dwarf Post helium-burning phase of stars with initial masses of less than eight times that of the Sun. Such stars are supported against gravitational collapse by degenerate electron pressure.

Wien's law If a blackbody radiator of temperature T emits a maximum energy flux at wavelength λ_{max}, then Wien's law dictates that the product λ_{max} T is a constant.

Appendix A:

A Homogeneous Star Model

The mass-luminosity relationship for a fully mixed, chemically homogeneous Sun-like star was described in Chapter 3. Indeed, Equation (3.12) indicates that the luminosity L is related to the mass M and the chemical composition via the relationship $L = L_0 \, \mu^{7.5} \, (1 + X)^{-1} \, M^5$, where $L_0 = L(X = X_0)$ is a constant, μ is the mean molecular weight and $0 \leq X \leq X_0$ is the hydrogen mass fraction of the stellar gas. When the mass fraction of the chemical elements other than hydrogen (X) and helium (Y) within a star are small (which corresponds to the condition $Z \approx 0$), the expression for the mean molecular weight simplifies[1] to $\mu(X) \approx 2(1 + X)^{-2}$. With this approximation for the mean molecular weight, the mass-luminosity relationship becomes

$$L(X) = L_0(1+X)^{-16} \, M^5 \qquad \text{(A.1)}$$

Equation (A.1) determines the luminosity of a fully mixed star of mass M and hydrogen mass fraction X.

In order to determine the effect of mass loss upon our model star, we assume that the mass loss rate is proportional to the star's luminosity and accordingly write,

$$\frac{\Delta M}{\Delta t} = N \, \frac{L}{c^2} \qquad \text{(A.2)}$$

where ΔM is the amount of mass lost by the star in the time interval Δt, N is a numerical parameter that can vary from zero (indicating no mass loss) to a value as high as several hundred, L is the luminosity, and c is the speed of light.

To make further progress Equation (A.2) needs to be converted into an expression that varies with the hydrogen mass fraction ΔX, rather than time Δt. This, however, is easy to do. Recall from Chapter 3 that the energy generated by the fusion reactions at the center of a star exactly compensates for the energy lost into space at its surface. Furthermore, to generate the energy that the star will eventually radiate into space, a certain amount of the hydrogen must be consumed. The amount of hydrogen consumed ΔX in the time interval Δt is determined, therefore, by the relationship

$$\frac{\Delta X}{\Delta t} = -\frac{L}{QM} \tag{A.3}$$

where Q is the energy liberated per kilogram of stellar material by nuclear fusion reactions, and the negative sign indicates that the hydrogen mass fraction decreases with time.

Now, combining Equations (A.2) and (A.3) the variation in the mass of the star can be expressed as

$$\frac{\Delta M}{M} = -N \frac{Q}{c^2} \Delta X \tag{A.4}$$

Equation (A.4) can now be integrated to reveal how the star's mass changes with decreasing X. Indeed, the mass of star decreases exponentially, and

$$m = \exp[k(x-1)] \tag{A.5}$$

where, the following short-hand notation has been introduced

$$m = \frac{M(X)}{M(X = X_0)} \quad x = \frac{1+X}{1+X_0} \tag{A.6}$$

and where the mass loss parameter is given by the expression

$$k = N \frac{Q}{c^2} (1 + X_0) \tag{A.7}$$

with $X_0 = 0.7$ being the initial hydrogen mass fraction, and where we have assumed that $N > 0$. If we also introduce the notation $l = L(X) / L(X = X_0)$, then the luminosity equation (A.1) becomes

$$l = x^{-16} m^5 \tag{A.8}$$

where m is given by Equation (A.5). It can now be seen that the higher the value of N, the greater the mass-loss rate [care of Equation (A.7)], and the lower the star's luminosity for all values of X. For very high mass-loss rates the luminosity can, in fact, be driven to values less than $L(X = X_0)$.

The value of the mass-loss parameter N need not be taken as a fixed quantity and it can certainly be allowed to vary. Indeed, by engineering the mass-loss rate appropriately a star can be made to evolve with a constant luminosity (this is mass-loss Scenario 1 described in Chapter 5). It can be shown[2] that the condition $L(X) = L(X = X_0) = $ constant, is achievable provided

$$N_1 = \left(\frac{16}{5}\right)\frac{c^2/Q}{(1+X)} = \frac{457}{(1+X)} \qquad (A.9)$$

Under the assumption that Earth's orbital radius increases in accordance with the conservation of angular momentum, the mass-loss rate must be adjusted so that the quantity $L(X) / d^2 = $ constant (this is mass-loss Scenario 2 described in Chapter 5). The latter evolutionary condition is achieved provided

$$N_2 = \left(\frac{16}{7}\right)\frac{c^2/Q}{(1+X)} = \frac{327}{(1+X)} \qquad (A.10)$$

If the mass-loss rate is assumed to be constant throughout the entire hydrogen-burning phase, then Equation (A.5) reveals that the ratio of the final mass to the initial mass will be

$$M_f/M_0 = \exp[-N(Q/c^2)X_0] \qquad (A.11)$$

where $Q / c^2 = 0.007$ is the energy liberated per kilogram of stellar material by the PP chain of fusion reactions.

Notes and References

1. From Equation (3.4) the mean molecular weight becomes $\mu(X) = 2 / (1 + 3X)$ when Z is assumed to be zero. However, if one looks at the approximation a little more closely, it turns out that with only a small error the equation for $\mu(X)$ simplifies to the analytically more convenient form used in this appendix – see M. Beech, A novel stellar model: 'a sacrifice before the lesser

shrine of plausibility.' *Astrophysics and Space Science*, **168**, 253–261 (1990).

2. This point is discussed in, Blue stragglers as indicators of extraterrestrial civilizations? *Earth, Moon and Planets*, **49**, 177–186 (1990).

Appendix B:

An Accreting Black Hole Model

The accretion-driven luminosity of a black hole is derived from Einstein's well-known mass-energy equivalence formula $E = M c^2$, where E is the energy, M is the mass and c is the speed of light. If a quantity of mass ΔM falls into a black hole in time Δt, liberating energy ΔE, then the accretion luminosity $L_{acc} = \Delta E / \Delta t$ will be

$$L_{acc} = f \left(\frac{\Delta M}{\Delta t} \right) c^2 \qquad (B.1)$$

where f is an efficiency factor accounting for the conversion of accreted material into energy, and where $\Delta M / \Delta t$ is the amount of material accreted into the black hole per unit time (i.e., the accretion rate). Now, the maximum accretion rate is typically taken to be that which produces a luminosity $L_{acc} = L_{Edd}$, where L_{Edd} is the so-called Eddington luminosity. The Eddington luminosity for an accreting black hole corresponds to the situation in which the radiation pressure on the infalling gas exactly balances the gravitational attraction of the black hole. For a black hole of mass M_{bh}, accreting at the Eddington rate the luminosity will be

$$L_{acc} = L_{Edd} = \frac{4\pi Gc}{0.02\,(1+X)} M_{bh} \qquad (B.2)$$

where it is assumed that the opacity is that due to electron scattering $\kappa = 0.02\,(1 + X)$ m^2/kg. The actual accretion rate of material into the black hole now becomes,

$$\frac{\Delta M_{bh}}{\Delta t} = (1-f) \left(\frac{\Delta M}{\Delta t} \right) \qquad (B.3)$$

Combining Equations (B.1), (B.2), with (B.3) and integrating reveals that the mass of the black hole increases exponentially with time.

Indeed, the black hole mass, a time t after emplacement within a star, will be

$$M_{bh}(t) = M_{bh}(0)\exp(t/\tau_{acc}) \tag{B.4}$$

where $M_{bh}(0)$ is the initial black hole mass, and the e-folding time, during which the black hole mass increases by a factor of $e = 2.718\ldots$, is

$$\tau_{acc} = \frac{0.02\,(1+X)c}{4\pi G}\,\frac{f}{1-f} \approx 3.4\text{x}10^8\,\frac{f}{1-f}\ (\text{yrs}) \tag{B.5}$$

where we have taken the hydrogen mass fraction in the core to be $X = 0.5$. For a typically assumed mass conversion efficiency of 10 percent ($f = 0.1$) the e-folding time is about 4×10^7 years. If the efficiency is as high as, say 25 percent then the e-folding time increases by a modest amount to about 10^8 years. In comparison to the main-sequence lifetime T_{MS} of the host star [given by Equation (2.1)], the time $T_{consume}$ for the black hole to fully consume the host star, is

$$\frac{T_{consume}}{T_{MS}} = 0.03\frac{f}{1-f}\,M_{star}^3\,\ln\left[\frac{M_{star}}{M_{bh}}\right] \tag{B.6}$$

where M_{star} and M_{bh} are the initial masses of the star and the black hole.

Index

Angular momentum, 124, 141, 156
Annis, James, 21, 46, 51
Anthropic Principle, 8, 18, 71, 110
Arkhipov, Alexey, 13–14
Asteroids individual
  2004 VD, 17
  (25143) Itokawa, 33
Aston, Francis, 25
Atomic bomb, 17, 199–200

Ball lightning, 189
Beckett, Margaret, 201
Bethe, Hans, 85
Birch, Paul, 58, 69, 126
Black dwarf, 49, 65, 89, 94
Black holes, 74, 90, 163, 171, 189, 221
Black smokers, 141
Blue Stragglers, 184–188
Bostrom, Nick, 197, 201, 206
Brown dwarf, 90, 109, 133, 144
Brown, Peter, 208

Carrying capacity, 206, 208
Carter, Brandon, 8, 9, 105, 111, 181
Cathcart, Richard, 162, 177, 178
Chandrasekhar limit, 40, 108, 111
Circumstellar disks, 184, 188, 191
CNO cycle, 108
CO_2-cycle, 54–55, 114, 139, 208
Comets
  impacts, 36–38
  Tempel 1, 38
  Wild 2, 35
Copernicus, Nicolaus, 7
COROT, 192
Cosmos-1, 129
Crab Nebula, 39
Criswell, David, 58, 69, 166–169
Croll-Milankovitch cycle, 53–54, 62

Darwin, Charles, 198
Derrida, Jacques, 197–198
Drake, Frank, 9, 18, 71, 136
Dyson sphere, 14, 16, 52–53, 66, 169, 182

Earth's orbit, 94, 123, 127–128, 156,
  158, 182
Eddington, Arthur, 23, 24, 73, 172, 221
Eddington luminosity, 179, 221
Enceledus, 19
Europa, 19, 115–117, 140
Extrasolar planets, 115, 125, 181–184
Extremophiles, 117, 122, 140–141

Fermi, Enrico, 2, 5, 16, 100
Fermi Paradox, 6, 47, 58, 137
Fermi problems, 17, 100
Fish stocks, 198
Fogg, Martyn, 58, 168, 178, 189, 195

Gaia, 56–57, 148, 176
Gamow, George, 86
Geminga, 41–42, 63
Gott, Richard III, 202
Gould, Stephen, 7, 18
GRBs, 44–47, 51
Greenhouse effect, 54, 109, 114,
  138, 140

Habitability zone, 113–115, 120–122
Haldane, J. B. S., 49
Hawking, Stephen, 189, 195, 205
Herschel, William, 94
Holmes, Sherlock, 99, 184
Hot Jupiters, 191–192, 195
HR diagram, 88–90, 91, 96, 159, 185
Hurricane Katrina, 27
Hypernovae, 44–47

Impact craters
  Chicxulub, 30, 61
  Wilkes Land, 37

Kardashev, Nicolai, 14, 16, 21,
    69, 178
Korycansky, Don, 127–128, 142

Lake Vostok (Antarctica), 116–117
Long Now Project, 201–202, 207
Lovelock, James, 56–57, 148, 176

M31 – Andromeda Galaxy, 47–48, 65
Main sequence lifetime, 25–26, 91, 135,
    144, 151, 161, 172
Malthus, Thomas, 59, 70
Mass extinction, 37, 45, 61
Mayer, Ernst, 18
Mean molecular weight, 77–78
Meteorite impacts, 28, 60
Mitalas, Romas, 60, 64, 176, 177
Modis cycle, 204, 208
Mount Pinatubo, 67

Nagasaki, 199
Nanotechnology, 200, 211
Near Earth Asteroid (NEA), 28, 33, 57, 61,
    128, 214
Neutron stars, 63, 90, 109

Obliquity, 143
O'Neill, Gerard, 47, 68–69,
    169, 203
Oort cloud, 34, 62, 170

Pauli exclusion principle, 101, 110
Pioneer spacecraft, 12
Planets
  Mercury, 92, 122–125, 203
  Venus, 53, 123–124, 129, 142, 203
  Earth, 53–57, 123, 133, 141, 143
  Mars, 53, 66, 203
  Jupiter, 121–122, 125, 131–133
  Saturn, 122, 125, 131
  Supramundane, 58, 69
  Neptune, 126
  Uranus, 126
Project Ozma, 18
Proton-Proton chain, 24, 85, 90, 108
Pulsars, 12, 63

Ramscoop (Bussard), 165–167,
    178, 191
Rees, Sir Martin, 110, 197
Reeves, Hubert, 1, 162, 177
Russel, Mary Doria, 20

Sagan, Carl, 6, 12
SETI, 13, 130, 191
Smith, Robert Connon, 106, 142, 194
Solar neutrino problem, 171, 179, 195
Solar sail, 32, 128–129, 130, 190
Solar siblings, 169–171
Solar wind, 155
Solar wrap, 173–175, 179, 182
Space elevator, 203, 208
Star lifting, 168–169, 191
Stars
  central pressure, 75,150, 154
  central temperature, 78,103, 150
  collisions between, 33–34, 48
  dynamical timescale, 73–75
  energy generation, 83–84, 86
  energy transport, 80–82
  evolution, 98, 133, 137, 142
  formation, 10, 48, 91, 100, 105, 134
  hydrostatic equilibrium,
  Kelvin-Helmholtz timescale, 84, 95
  mass-luminosity law, 87–88, 144,
      154, 217
  mean free path, 79, 82–83
  mean molecular weight, 77, 87, 95–96,
      99, 154
  opacity, 79, 83, 98, 152–153
  photon diffusion time, 78–80,
      83, 152
  red giant stage, 93, 96, 109–110
  spectral type, 71, 106
Stars – individual
  16 Cygnus B, 115, 121
  47 Ursae Majoris, 114, 121
  51 Pegasi, 191
  55 Cancri, 184
  70 Virginis, 183
  CW Leonis, 131–132
  Cygnus X-1, 163
  Epsilon Eridani, 18
  Eta Carina, 42–43
  GD356, 133
  Gliese 581, 118
  Gliese 710, 35–36, 168
  HD4308, 183
  HD33487, 36
  HD69830, 115, 121

HD99109, 183
HD 158576, 36
HD179939, 36
HD183263, 121
HD190360, 183
HD209458, 196
HD210277, 184
HD213240, 183
HD216435, 183
IK Pegasus, 40, 42–43
OGLE-TR-56, 125
Omicron Ceti (Mira), 93
Proxima Centauri, 33, 34, 35,
    47, 63
Rho Corona Borealis, 184
Tau Ceti, 18
V838 Monocerotis, 119–121, 140
WD0137-349, 133–134
Steel, Duncan, 63
Stellifying Jupiter, 157
Stranglets, 50
Sun
    Evolution, 90–95, 124, 142
    Mass loss, 93, 148–150, 155, 217
    Mixing, 148–150, 162, 217
    Rotation, 150, 152
    Structure, 25, 174
Sunshield, 46, 55, 65, 160

Supernova
    Type I, 41–43, 63, 90
    Type II, 42, 48, 63, 90
Synchronous rotation, 118–119

Terraforming, 53–57, 129, 190, 201
Terrestrial Planet Finder (TPF), 192
Tidal heating, 116
Tolkien, J. R. R., 211
Torino Scale, 31–33
Triple alpha reaction, 84, 108, 110
Tunguska event, 29–30, 60, 203

UV problem, 46, 159–160

Vonnegut, Kurt, 211
von Neumann, John, 11, 19

Water worlds, 126, 142, 191
Wells, Herbert George, 209
White Dwarf, 40, 81, 89–90, 94, 96,
    132–133
Wolfram, Stephen, 13
Wolf-Rayet stars, 64

X-ray emission, 163, 190

Printed in the United States of America